MANAGING WATER RESOURCES: GUIDE TO IRRIGATION & HYDROLOGY

By:

Almas Qureshi
Assistant Professor
Acropolis Institute of Technology & Research,
Indore

Preface

Water is one of the most critical resources on our planet, fundamental to life, agriculture, and industry. As our global population continues to grow, the demand for water has surged, making the effective management of this precious resource more important than ever. This book, **"Managing Water Resources: A Guide to Irrigation and Hydrology,"** is designed to provide comprehensive insights into the principles and practices of water resource management, specifically tailored for students pursuing a **Diploma in civil engineering at Rajiv Gandhi Proudyogiki Vishwavidyalaya (RGPV)** and other similar institutions.

The field of irrigation and hydrology encompasses a vast array of knowledge, from the understanding of water cycles and hydrological processes to the practical aspects of designing and managing irrigation systems. This guide aims to bridge the gap between theoretical concepts and real-world applications, equipping students with the skills necessary to address the challenges of water management in the 21st century.

Throughout this book, we have strives to present the material in a clear and accessible manner, with numerous illustrations, examples, and case studies to enhance understanding. Each chapter is structured to build on the previous ones, gradually deepening the reader's knowledge and competency in managing water resources effectively.

We begin with an overview of the fundamentals of hydrology, exploring the water cycle, precipitation, and runoff processes. From there, we delve into the principles of irrigation, discussing various methods, technologies, and best practices for efficient water use in agriculture.

This book is not only a textbook but also a practical guide that students can refer to throughout their academic and professional careers.

Finally, we dedicate this book to the students who are passionate about making a difference in the field of water resources management. Your commitment to learning and improving our world is what drives progress and innovation in this vital field.

With best wishes for your success,

Almas Qureshi   Indore

30-06-2024

Contents

CHAPTER 1 Hydrology, Hydrological Cycle 1
    Hydrology..1
    Hydrological Cycle..2

CHAPTER 2 Precipitation 4
    2.1 Precipitation..4

CHAPTER 3 Measurement of Rainfall..7
    3.1 Rainfall..7
    3.2 Measurement of Average Rainfall over a basin................13

CHAPTER 4 Crop Water Requirement..16
    4.1 Duty..16
    4.2 Delta...19

CHAPTER 5 Methods of Irrigation..21
    5.1 Methods of Irrigation...21
    5.2 Survey for Irrigation Project..23
    5.3 Data Collection for Irrigation Project..............................24
    5.4 Steps in conducting Survey..24
    5.5 Control Level in reservoir..25

CHAPTER 6 Dams and its Classification...26
    6.1 Dam...26

CHAPTER 7 Components of Earthen Dam 28
    7.1 Cross-Section of Earthen Dam...28

CHAPTER 8 Gravity Dam 30
    8.1 Gravity Dam..30

CHAPTER 9 Spillways 39
    9.1 Spillways..39

CHAPTER 10 Bandhara Irrigation 41
    10.1 Bandhara Irrigation..41

CHAPTER 11 Percolation Tank 44
    11.1 Percolation Tank...44

CHAPTER 12 Lift Irrigation Scheme 46
    12.1 Lift Irrigation..46

CHAPTER 13 Drift & Sprinkle Irrigation Scheme 48
    13.1 Drift Irrigation..48
    13.2 Sprinkle Irrigation..49

CHAPTER 14 Well Irrigation 51
    14.1 Well Irrigation..51

CHAPTER 15 Weirs 56
    15.1 Weirs..56

CHAPTER 16 Barrages 59
    16.1 Barrages.. 59

CHAPTER 17 Diversion Head works 62
    17.1 Diversion Head work.. 62

CHAPTER 18 Canals 64
    18.1 Canals.. 64
    18.2 Design of most economical canal section............................. 65
    18.3 Cross Drainage work... 67
    18.4 Aqueduct... 67
    18.5 Syphon Aqueduct... 68
    18.6 Super passage... 68
    18.7 Level Crossing... 69

CHAPTER 19 Canal Regulator 70
    19.2 Canal Escape... 72
    19.3 Canal Outlet.. 73
    19.4 Cross Regulator... 75
    19.5 Head Regulator... 75

Chapter 1: Hydrology, Hydrolological Cycle

1.1 HYDROLOGY:

1.1.1 Definition:
Definition: Hydrology is the scientific study of water in the environment.
Focus Areas: It looks at the distribution, movement, and properties of water on Earth.
Components: Includes the study of rivers, lakes, groundwater, glaciers, and precipitation (like rain and snow).

1.1.2 : History of Hydrology:
Ancient Civilizations:
Early Beginnings: The study of water started thousands of years ago.
Egyptians and Mesopotamians: Ancient Egyptians and Mesopotamians built irrigation systems and studied river behaviors for agriculture.

Greek Contributions:
Hippocrates: Around 400 BC, Hippocrates discussed the health effects of water.
Aristotle: Around the same time, Aristotle studied water cycles and evaporation.

Medieval Period:
Islamic Scholars: In the medieval period, Islamic scholars like Al-Biruni (973-1048 AD) made significant contributions to hydrology by studying the hydrological cycle.

Renaissance to 18th Century:
Leonardo da Vinci: During the Renaissance, Leonardo da Vinci (1452-1519) studied water flow and sediment transport.
Pierre Perrault and Edme Mariotte: In the 17th century, Pierre Perrault and Edme Mariotte conducted experiments on rainfall and river flow.

19th Century:
Advances in Measurement: This period saw improvements in measuring river discharge, rainfall, and groundwater.
Robert Manning: In 1891, Robert Manning developed the Manning formula to calculate water flow in channels.

20th Century to Present:
Modern Hydrology: The 20th century brought more scientific methods and technology, such as remote sensing and computer modeling.
Integrated Water Management: Today, hydrology also focuses on managing water resources sustainably.

1.2 HYDROLOGICAL CYCLE:

1.2.1 Definition:
Definition: The hydrological cycle, also known as the water cycle, is the continuous movement of water on, above, and below the Earth's surface.

Importance: It is essential for maintaining life on Earth, influencing weather and climate, and supporting ecosystems.

1.2.2 Components of Hydrological Cycle:

1 Evaporation:
Process: Water from oceans, rivers, lakes, and other surfaces turns into water vapor due to heat from the sun.
Result: This water vapor rises into the atmosphere.

2 Transpiration:
Process: Plants absorb water from the soil and release it as water vapor through their leaves.
Result: This adds to the moisture in the atmosphere.

3 Condensation:
Process: Water vapor in the air cools down and changes back into liquid droplets.
Result: These droplets come together to form clouds.

4 Precipitation:
- *Process:* When clouds become heavy with water droplets, they release this water as rain, snow, sleet, or hail.
- *Result:* This water falls back to Earth's surface.

Infiltration:
Process: Some of the water from precipitation soaks into the ground.
Result: This water replenishes underground aquifers and becomes groundwater.

Runoff:
Process: Water that does not infiltrate the ground flows over the surface and collects in rivers, lakes, and oceans.
Result: This water eventually returns to the sea.

Percolation:
Process: Water moves downward through the soil and rocks.
Result: This water reaches deeper underground layers, replenishing aquifers.

Groundwater Flow:
Process: Groundwater moves slowly through soil and rocks.
Result: It may eventually return to the surface through springs or flow into rivers, lakes, or oceans.

Sublimation:
Process: In cold climates, ice and snow can directly turn into water vapor without melting first.
Result: This adds water vapor to the atmosphere.

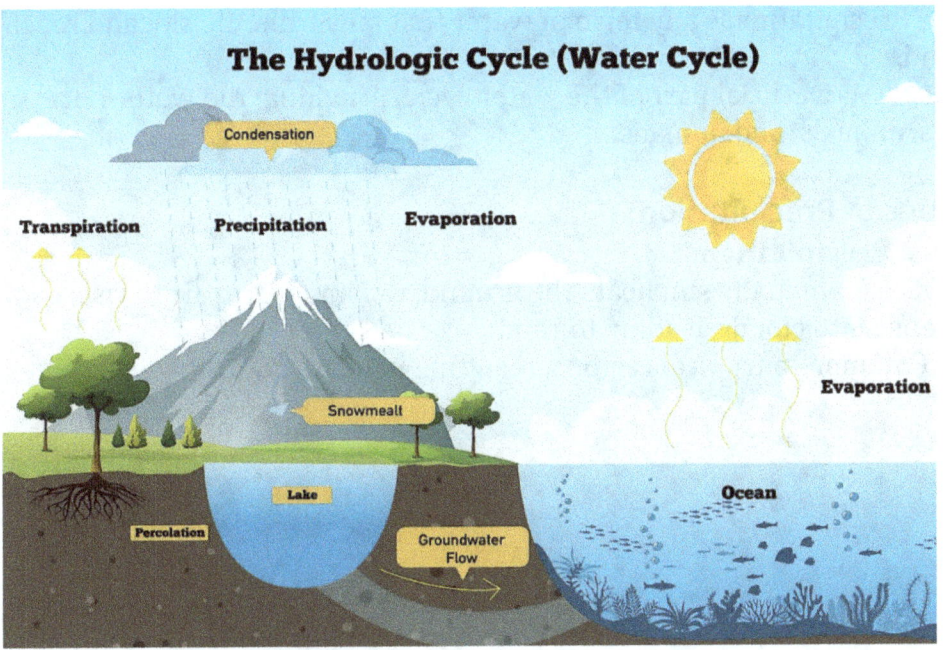

Figure: Hydrological Cycle

Chapter 2: Precipitation

2.1 PRECIPITATION:

2.1.1 Definition:

Definition: Precipitation is any form of water that falls from the sky and reaches the ground.

Importance: It is a crucial part of the water cycle, providing the water necessary for plants, animals, and humans.

2.1.2 Types of Precipitation:

Convective Precipitation:

Process: Occurs when the sun heats the ground, causing warm air to rise, cool, and condense into clouds, leading to rain.

Example: Common in tropical regions, resulting in heavy rain showers.

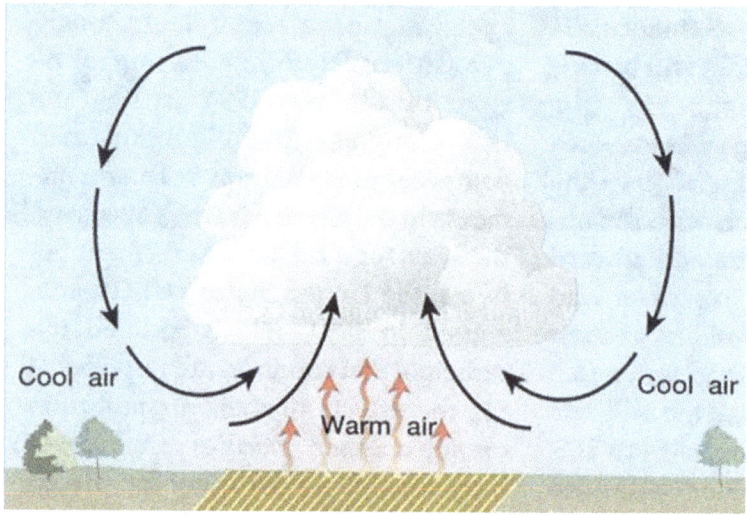

Figure: Convective Precipitation

Orographic Precipitation:

Process: Happens when moist air is forced to rise over mountains, cooling and condensing into clouds and precipitation.

Example: Rain or snow on the windward side of mountains.

Figure: Orographic Precipitation

Frontal Precipitation:
Process: Occurs when a warm air mass meets a cold air mass, causing the warm air to rise, cool, and form precipitation.
Example: Common in mid-latitude regions, leading to various types of precipitation like rain or snow.

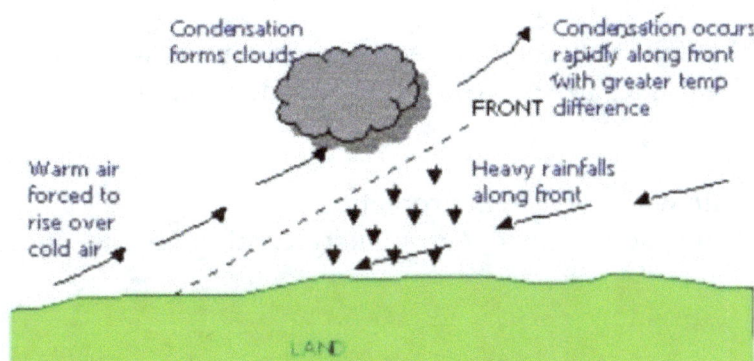

Figure: Frontal Precipitation

Cyclonic (or Depressional) Precipitation:
Process: Associated with low-pressure systems where air spirals inward and upward, cooling and causing precipitation.
Example: Typical in storm systems and cyclones.

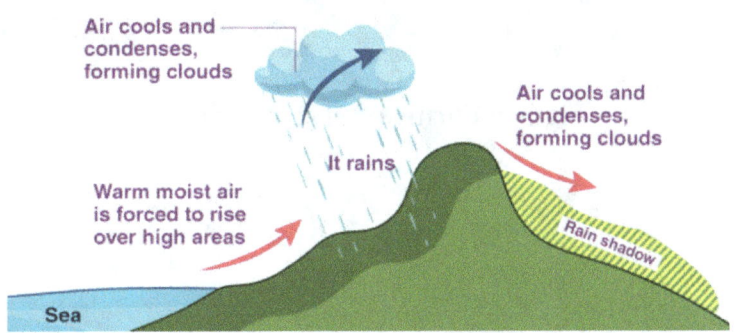

Figure: Cyclonic Precipitation

2.1.3 Forms of Precipitation:

Rain:
Definition: Liquid water droplets that fall when temperatures are above freezing.
Characteristics: Can vary in intensity from light drizzle to heavy downpours.

Snow:
Definition: Ice crystals that form when temperatures are below freezing.
Characteristics: Snowflakes have unique shapes and accumulate on the ground.

Sleet:
Definition: Small ice pellets that form when raindrops freeze before hitting the ground.
Characteristics: Can make surfaces slippery and hazardous.

Hail:
Definition: Balls or lumps of ice that form in strong thunderstorms with intense updrafts.
Characteristics: Hailstones can range in size from small peas to large golf balls or even bigger.

Freezing Rain:
Definition: Rain that falls as liquid but freezes upon contact with cold surfaces.
Characteristics: Creates a glaze of ice, leading to dangerous conditions.

Drizzle:
Definition: Light rain with very small droplets.
Characteristics: Often occurs in foggy or misty conditions and doesn't accumulate much

Figure: Forms of Precipitation

Chapter 3: Measurement of Rainfall

3.1 RAINFALL:

3.1.1 Definition:
Definition: Rainfall is the amount of rain that falls over a specific area in a certain period.
Measurement: It is measured in millimeters (mm) or inches.

3.1.2 Types of Raingauge:
3.1.2.1 Rain Gauge
Definition: A rain gauge is a tool used to measure the amount of rainfall.
Purpose: Helps in studying weather patterns, climate, and managing water resources.

3.1.2.1 Types of Rain Gauge

NON-AUTOMATIC RAIN GAUGE:
Description: Simple, manual devices that require human intervention to record data.
Example: Standard rain gauge.

A.1 Standard Rain Gauge:

Components:

Funnel: Directs rainwater into the measuring cylinder.

Measuring Cylinder: Graduated cylinder to measure rainfall.

Specifications:

Material: Usually made of clear plastic or glass.

Height: About 12-24 inches (30-60 cm).

Diameter: Funnel opening around 4-8 inches (10-20 cm).

Graduations: Marked in millimeters or inches for accurate measurement.

Usage:

Placement: Installed on a flat, open ground away from obstructions.

Reading: Rainwater is manually measured by reading the level in the cylinder.

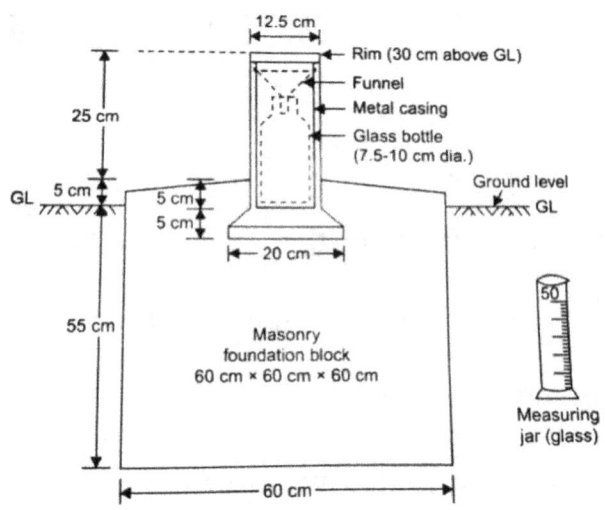

Figure: Symon's Rain Gauge

AUTOMATIC RAIN GAUGE:

Description: Devices that measure and record rainfall automatically without human intervention.

Types: Weighing bucket, tipping bucket, and float type.

B.1 Weighing Bucket Rain Gauge:

- **Components**:

    Bucket: Collects rainwater.

    Weighing Mechanism: Measures the weight of the collected water.

    Recording System: Records the weight data.

- **Specifications**:

    Material: Typically stainless steel or durable plastic.

    Capacity: Varies, usually up to 12 inches (300 mm) of rainfall.

    Accuracy: High, typically within 0.1 mm.

    Dimensions:

    Height: About 20-30 inches (50-75 cm).

    Diameter: Bucket opening around 8-12 inches (20-30 cm).

- **Usage**:

    Advantages: Measures both liquid and solid precipitation.

    Data Output: Continuous data logging and real-time monitoring.

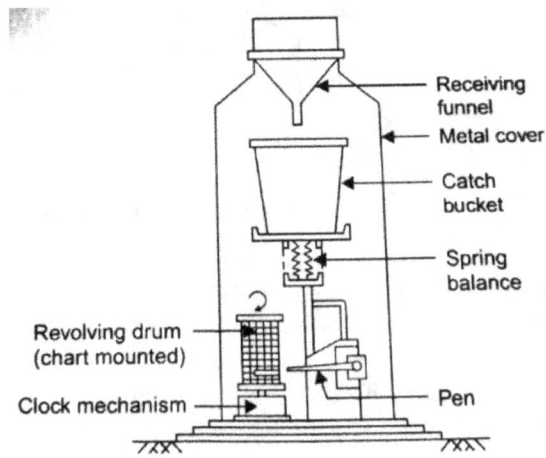

Figure: Weighing Bucket Rain Gauge

B.2 Tipping Bucket Rain Gauge:

- **Components**:

    Tipping Buckets: Two small buckets on a pivot.

    Funnel: Directs rain into the buckets.

    Counter/Recorder: Records each tip of the bucket.

- **Specifications**:

    - **Material**: Typically made of plastic or metal.
    - **Capacity**: Each bucket tip usually represents 0.01 inch (0.2 mm) of rain.
    - **Accuracy**: High, suitable for most meteorological needs.
    - **Dimensions**:

        Height: About 12-24 inches (30-60 cm).

        Diameter: Funnel opening around 8-12 inches (20-30 cm).

- **Usage**:

    Advantages: Provides real-time data, widely used in weather stations.

    Data Output: Digital output for data logging and remote monitoring.

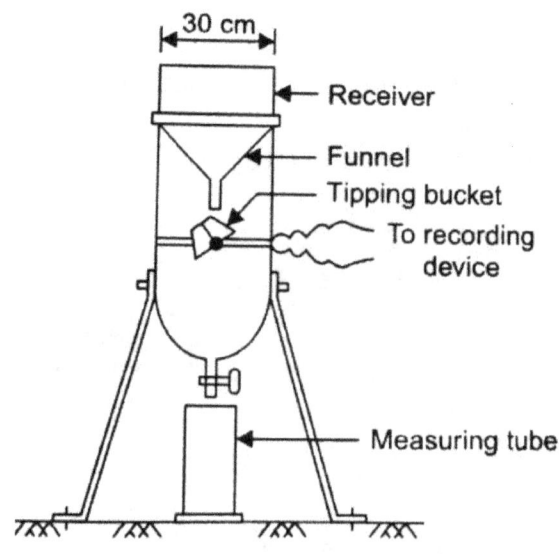

Figure: Tipping Bucket Rain Gauge

B.2 Float Type Rain Gauge:

- **Components**:

    Float: Positioned in a collecting container.

    Container: Collects rainwater.

    Recording Device: Tracks the movement of the float.

- **Specifications**:

    Material: Container usually made of metal or plastic.

    Capacity: Typically up to 12 inches (300 mm) of rainfall.

    Accuracy: High, within 0.1 mm.

    Dimensions:

    Height: About 20-30 inches (50-75 cm).

    Diameter: Container opening around 8-12 inches (20-30 cm).

- **Usage**:

    Advantages: Continuous recording, suitable for remote areas.

    Data Output: Continuous data logging and real-time monitoring.

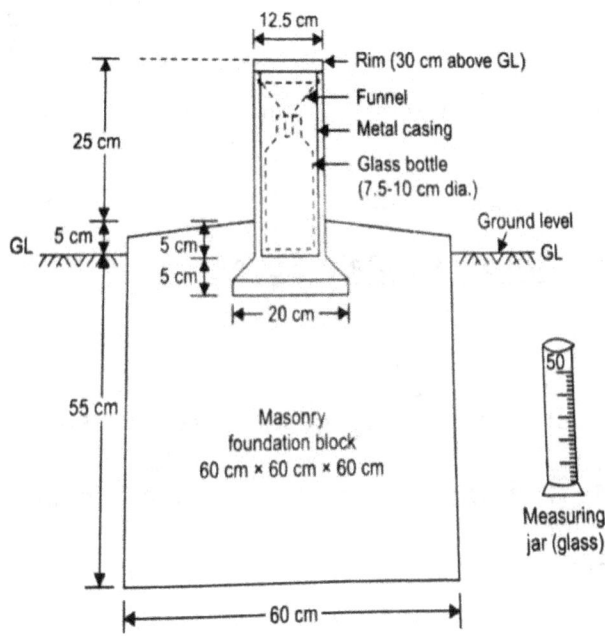

Figure: Float Type Rain Gauge

3.1.3 Errors in measurement of rainfall:

3.1.3.1 General Errors:

Splash-In and Splash-Out:

Splash-In: Water from outside the gauge can splash into it, causing an overestimation.

Splash-Out: Rainwater already in the gauge can splash out, causing an underestimation.

Wind Effects:

Strong Winds: Can blow rain past the gauge or cause turbulence, leading to inaccurate readings.

Sheltering: Placement of the gauge near buildings or trees can block rain, leading to underestimation.

Evaporation:

High Temperatures: Can cause collected water to evaporate before it is measured, leading to underestimation.

Delay in Reading: Manual gauges are more susceptible to this error if not checked promptly.

Obstructions:

Debris: Leaves, dirt, or insects can block the funnel, reducing the amount of water collected.

Ice and Snow: Can block the gauge, preventing accurate measurement until they melt.

3.1.3.2 Errors in Non-Automatic Rain Gauge:

Human Error:

Reading Mistakes: Incorrectly reading the measurement scale can cause errors.

Recording Errors: Misreading the data can lead to inaccuracies.

Timing of Measurement:

Irregular Checks: Not checking the gauge regularly can lead to missing rainfall events.

Delay: Water can evaporate or spill if not measured promptly.

3.1.3.2 Errors in Automatic Rain Gauge:

1. Weighing Bucket Rain Gauge:

Calibration Errors: Inaccurate calibration of the weighing mechanism can cause errors.

Sensitivity: Very small amounts of precipitation might not be detected accurately.

Mechanical Issues: Malfunctioning scales can provide incorrect readings.

Tipping Bucket Rain Gauge:

Bucket Sticking: If buckets do not tip correctly, it can cause underestimation.

Resolution: Each tip represents a fixed amount of rain; very light rain might not be recorded accurately.

Overflow: Heavy rain can cause the gauge to miss some tips if buckets fill too quickly.

Float Type Rain Gauge:

Float Jamming: The float can get stuck, causing incorrect measurements.

Calibration: The recording device must be accurately calibrated to track float movement.

Sediment Accumulation: Dirt and debris can affect float movement and accuracy.

3.2 MEASUREMENT OF AVERAGE RAINFALL OVER A BASIN:

ARITHMETIC MEAN METHOD:

1. **Description:**

   - **Simple Method**: Calculates the average of rainfall recorded at different stations.
   - **Formula**:

   $$\text{Average Rainfall} = \frac{\sum P}{n}$$

   - P: Rainfall at each station
   - n: Number of stations

2. **Example:**

   - **Stations**: 3 stations with rainfall of 10 mm, 15 mm, and 20 mm.
   - **Calculation**:

   $$\text{Average Rainfall} = \frac{10 + 15 + 20}{3} = \frac{45}{3} = 15 \text{ mm}$$

   - **Result**: The average rainfall is 15 mm.

THEISSEN POLYGON METHOD:

1. **Description:**

   - **Weighted Average Method**: Assigns different weights to stations based on their area of influence.
   - **Steps:**
     - Draw polygons around each station such that each point within a polygon is closer to that station than any other.
     - Calculate the area of each polygon.
   - **Formula:**

   $$\text{Average Rainfall} = \frac{\sum (P_i \times A_i)}{\sum A_i}$$

   - P_i: Rainfall at station i

2. **Example:**

   - **Stations and Areas**: 3 stations with rainfall and areas:
     - Station 1: 10 mm, 20 km²
     - Station 2: 15 mm, 30 km²
     - Station 3: 20 mm, 50 km²
   - **Calculation:**

   $$\text{Average Rainfall} = \frac{(10 \times 20) + (15 \times 30) + (20 \times 50)}{20 + 30 + 50} = \frac{200 + 450 + 1000}{100}$$

   - **Result**: The average rainfall is 16.5 mm.

ISOHYTEL METHOD:

1. **Description:**

   - **Contour Method:** Uses contour lines (isohyets) to connect points of equal rainfall.
   - **Steps:**
     - Draw isohyets on a map.
     - Calculate the area between each pair of isohyets.
     - Assume rainfall between isohyets is the average of the isohyets.
   - **Formula:**

   $$\text{Average Rainfall} = \frac{\sum(P_i \times A_i)}{\sum A_i}$$

   - P_i: Average rainfall between isohyets i
   - A_i: Area between isohyets i

2. **Example:**

   - **Isohyets and Areas:** 3 isohyets with areas and rainfall:
     - Isohyet 1 (10 mm to 15 mm): Area = 20 km², Average rainfall = 12.5 mm
     - Isohyet 2 (15 mm to 20 mm): Area = 30 km², Average rainfall = 17.5 mm
     - Isohyet 3 (20 mm to 25 mm): Area = 50 km², Average rainfall = 22.5 mm
   - **Calculation:**

   $$\text{Average Rainfall} = \frac{(12.5 \times 20) + (17.5 \times 30) + (22.5 \times 50)}{20 + 30 + 50} = \frac{250 + 525 + 11}{100}$$

   - **Result:** The average rainfall is 19 mm.

Chapter 4: Crop Water Requirement

4.1 DUTY:

4.1.1 Definition:

Definition: Duty refers to the relationship between the amount of water supplied to a crop and the area of land that can be irrigated with that water over a specific period.

Expression: Typically expressed in hectares per cubic meter (ha/m³) or acres per cubic foot (ac/ft).

Importance: It is a critical factor in irrigation planning, ensuring that crops receive the necessary water for optimal growth without wastage.

4.1.2 Components of Duty:

Irrigation Period:

Time Frame: The period during which the irrigation water is applied to the crops, usually a growing season or a specific crop period.

Irrigated Area:

Land Area: The total area of land that receives irrigation water.

Volume of Water:

Water Quantity: The total volume of water used to irrigate the specified area.

4.1.3 Formula of Duty:

1. Formula:

$$\text{Duty} = \frac{\text{Area Irrigated (A)}}{\text{Volume of Water (V)}}$$

- A: Area irrigated (typically in hectares or acres)
- V: Volume of water supplied (typically in cubic meters or cubic feet)

Example Calculation

1. **Example:**

   - **Given:**
     - Area irrigated = 100 hectares
     - Volume of water supplied = 10,000 cubic meters

   - **Calculation:**

   $$\text{Duty} = \frac{100 \text{ hectares}}{10,000 \text{ cubic meters}} = 0.01 \text{ hectares per cubic meter } (\text{ha/m}^3)$$

   - **Result:** The duty is 0.01 ha/m³, meaning 1 cubic meter of water can irrigate 0.01 hectares of land.

4.1.4 Importance of Duty in Crop water requirement:

1. **Efficient Water Use:**

   - **Optimization:** Helps in optimizing the use of available water resources for irrigation.
   - **Conservation:** Ensures minimal wastage and effective distribution of water.

2. **Crop Planning:**

   - **Scheduling:** Assists in scheduling irrigation based on crop water requirements and available water.
   - **Yield Improvement:** Ensures that crops receive the necessary water for healthy growth and maximum yield.

3. **Irrigation Management:**

   - **Infrastructure Design:** Helps in designing irrigation systems and infrastructure, such as canals and reservoirs.
   - **Resource Allocation:** Aids in the allocation of water resources among different crops and Water use.

4.1.5 Factors Affecting Duty:

A. Methods & System of Irrigation:
How farmers water their fields affects duty.
Some methods like drip irrigation use water more efficiently than others like flood irrigation.
The system of irrigation, like whether it's controlled by canals or pumps, also impacts duty.

B. Mode of Applying Water to Crops:
Different ways of giving water to plants affect how much water they need. For example, spraying water might need less than flooding fields.

C. Methods of Cultivation:
How farmers plant and take care of crops can change how much water those crops need.
Techniques like no-till farming can reduce water loss.

D. Time and Frequency of Tilling:
Tilling, or digging up soil, affects how much water stays in the ground.
Doing it at certain times or not too often can help keep water where plants can use it.

E. Types of Crop:
Some plants are thirstier than others.
Rice, for example, needs a lot of water, while cacti need very little.

F. Base Period of Crop:
This is about when crops need the most water.
Some need it all at once, while others need it spread out over time.

G. Climatic Conditions of Area:
Hotter places usually need more water for crops because it evaporates faster. Rainfall also matters – if it's not enough, farmers need to water their crops more.

H. Quality of Water:
Not all water is good for plants.
If it's too salty or dirty, it can harm crops or make them need more water.

I. Method of Assessment of Irrigation Method:
How farmers check if their watering methods are working well affects how much water they use.
If they can measure it accurately, they can adjust better.

J. Canal Conditions:
Canals that carry water to fields need to be in good shape.
Leaks or blockages can waste water or stop it from reaching crops.

K. Character of Soil and Sub-soil of Canal:

Different soils hold water differently.
Sandy soil lets water pass through quickly, while clay soil keeps it in place.

L. Character of Soil and Sub-soil of Irrigation Fields:
Just like with canals, the type of soil in fields matters.
Some soak up water well, while others let it run off.

4.2 DELTA:

4.2.1 Definition:
Definition: Delta is the total depth of water required by a crop during its entire growing season.
Expression: Typically expressed in centimeters (cm) or inches (in).
Importance: It represents the total water demand of a crop from planting to harvest, ensuring the crop's water needs are met for optimal growth and yield.

4.2.2 Components of Delta:
Growing Season:

Time Frame: The period from planting to harvest during which the crop grows and matures.

Total Water Requirement:

Volume of Water: The cumulative amount of water needed by the crop throughout its growing season.

4.2.3 Formulae of Delta:
1. Formula:

$$\Delta = \frac{V}{A}$$

- Δ: Delta (total depth of water in cm or inches)
- V: Total volume of water supplied (typically in cubic meters or acre-feet)
- A: Area irrigated (typically in hectares or acres)

4.2.4 Example Calculation:

1. **Example:**

   - **Given:**
     - Total volume of water supplied = 30,000 cubic meters
     - Area irrigated = 10 hectares
   - **Calculation:**
     $$\Delta = \frac{30,000 \text{ cubic meters}}{10 \text{ hectares}} = 3,000 \text{ cubic meters per hectare}$$
   - **Convert to Depth:**
     - 1 hectare = 10,000 square meters
     - 3,000 cubic meters per hectare = 3,000 / 10,000 = 0.3 meters = 30 cm
   - **Result:** The delta is 30 cm, meaning the crop requires a total of 30 cm of water over its growing season.

4.2.5 Importance of Delta in Crop water requirement:

1 Water Planning:

Irrigation Scheduling: Helps in planning the irrigation schedule to meet the crop's water needs at different growth stages.

Resource Allocation: Aids in the efficient allocation of water resources among different crops and fields.

2 Crop Management:

Growth Optimization: Ensures that crops receive the necessary water for healthy growth, reducing water stress and improving yield.

Yield Improvement: Adequate water supply throughout the growing season can lead to better crop yields.

3 Infrastructure Design:

Irrigation Systems: Helps in designing and managing irrigation systems to deliver the required amount of water efficiently.

Storage Requirements: Assists in determining the storage capacity needed for reservoirs or other water storage facilities.

Chapter 5: Methods of Irrigation

5.1 Methods of Irrigation:

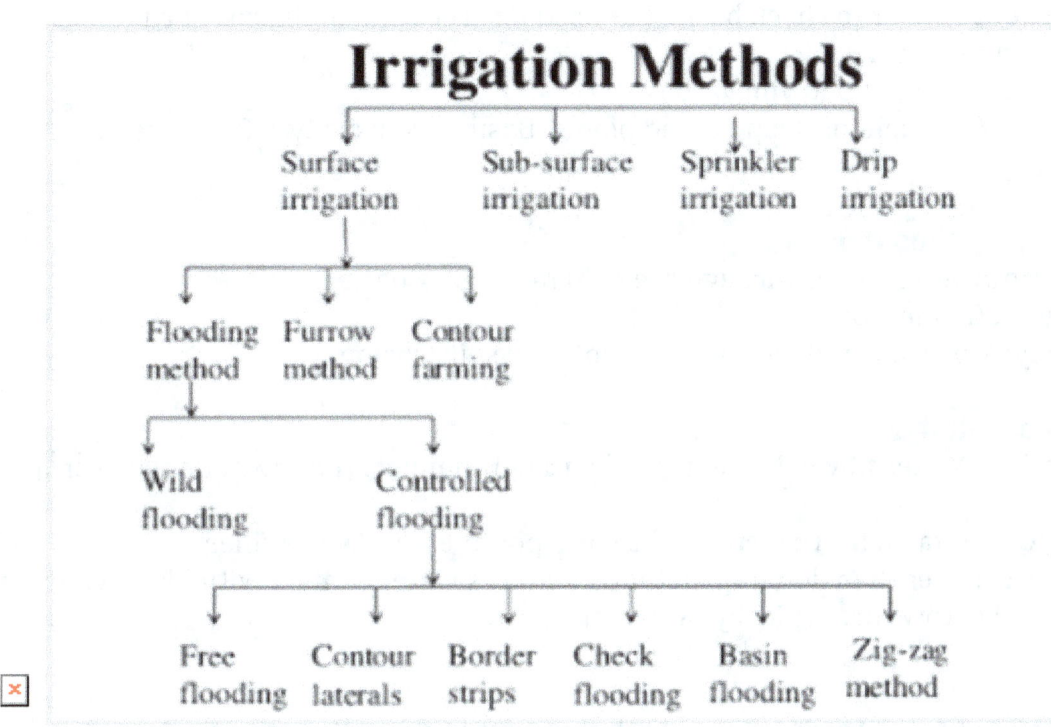

A. SURFACE IRRIGATION
1. Flooding

1.1 Wild Flooding (Free Flooding)
Description: Water is released onto the field without any specific channels or controls.
Land Type: Suitable for uneven or irregular fields.
Efficiency: Low, as water spreads unevenly leading to potential water logging and wastage.

1.2 Contour Lateral Flooding
Description: Water is directed along the natural contour lines of the field.
Land Type: Sloped or hilly lands.
Efficiency: Moderate, helps in reducing soil erosion and water runoff.

1.3 Border Strip Flooding
Description: Field is divided into long, narrow strips separated by small ridges.
Land Type: Flat lands with slight slopes.
Efficiency: Higher than wild flooding, allows better water distribution. Typical strip dimensions are 10-20 meters wide and up to 400 meters long.

1.4 Check Basin Flooding
Description: Field is divided into small rectangular basins with ridges around them.
Land Type: Level fields, often used for rice cultivation.
Efficiency: High, allows precise water control within each basin. Basin size typically ranges from 3x3 to 10x10 meters.

1.5 Basin Flooding
Description: Similar to check basin, but used for trees and orchards. Each tree is surrounded by a basin.
Land Type: Orchards and vineyards.
Efficiency: Very high for deep-rooted plants. Basins are usually 1-3 meters in diameter.

1.6 Zigzag Method of Flooding
Description: Water flows through the field in a zigzag pattern.
Land Type: Gentle slopes.
Efficiency: Moderate, reduces water runoff and soil erosion.

1.7 Furrow Method
Description: Water flows through small parallel channels (furrows) between crop rows.
Land Type: Suitable for row crops like corn, potatoes, and vegetables.
Efficiency: Higher than flooding methods, reduces water contact with plant leaves and stems. Furrows are typically 30-60 cm apart.

1.8 Contour Farming
Description: Crops are planted along the contour lines of the land, which are level curves across a slope.
Land Type: Hilly or sloped terrains.
Efficiency: High, as it helps in water retention and reduces soil erosion.

B.SUB-SURFACE IRRIGATION
Description: Water is delivered directly to the root zone below the soil surface using a network of buried pipes or tubes.
Installation: Involves laying pipes 15-90 cm below the surface, depending on crop type.
Efficiency: Very high, reduces water loss due to evaporation and deep percolation.
Cost: High initial setup cost due to pipe installation and maintenance requirements.
Usage: Ideal for high-value crops like vegetables and fruits, as well as landscapes and lawns.

Sprinkler Irrigation
Description: Water is sprayed through the air using a system of pipes, pumps, and sprinklers, mimicking natural rainfall.

System Types:
1 Fixed Systems: Sprinklers are permanently installed in the field.
2 Portable Systems: Sprinklers can be moved around the field.
3 Center Pivot Systems: A rotating arm sprays water as it moves around a pivot point, covering circular areas.

Efficiency: High, provides uniform water distribution and can be used on various terrains. Reduces water runoff and erosion.

Typical Applications: Used for a wide range of crops including cereals, vegetables, and fruit crops. Suitable for fields of various shapes and sizes.

Water Application Rates: Typically ranges from 5 to 15 mm per hour.

5.2 Survey for Irrigation Project:

3.1 Purpose of Survey
Objective: Determine the feasibility and design the layout of an irrigation project.
Outcome: Detailed information about the land, water sources, and environmental conditions.

3.2 Types of Surveys

A. Topographical Survey
Purpose: Map the physical features of the land.
Tools: Theodolites, Total Stations, GPS.
Data Collected: Elevations, slopes, natural features (rivers, hills), man-made features (roads, buildings).

Hydrological Survey
Purpose: Understand water availability and flow patterns.
Tools: Flow meters, rain gauges, satellite data.
Data Collected: Rainfall data, river flow rates, groundwater levels, existing irrigation systems.

Soil Survey
Purpose: Assess soil characteristics and suitability for irrigation.
Tools: Soil sampling kits, augers, lab analysis.
Data Collected: Soil texture, composition, pH levels, fertility, drainage capacity.

Socio-Economic Survey
Purpose: Understand the needs and conditions of the local community.
Tools: Interviews, questionnaires, demographic data.
Data Collected: Population density, land ownership, crop patterns, economic activities.

5.3 Data Collection for Irrigation Project:

Topographical Data
Elevation: Measures the height of the land above sea level.
Slope: The steepness of the land, crucial for determining water flow.
Contour Lines: Lines on a map that connect points of equal elevation, used to design efficient irrigation channels.

Hydrological Data
Rainfall Data:
Annual Rainfall: Total amount of rainfall in a year.
Seasonal Distribution: Rainfall patterns during different seasons.

River Flow Rates
Discharge: The volume of water flowing through a river, usually measured in cubic meters per second (m^3/s).
Flow Variability: Changes in river flow during dry and wet seasons.

Groundwater Levels
Water Table Depth: The level at which soil pore spaces are saturated with water.
Aquifer Characteristics: Properties of underground layers that store water.

Soil Data
Texture: Proportion of sand, silt, and clay in the soil.
Structure: The arrangement of soil particles into aggregates.
Fertility: Nutrient content of the soil, essential for crop growth.
Drainage: Soil's ability to allow water to pass through it.

Socio-Economic Data
Population Density: Number of people living per unit area.
Land Ownership: Distribution and size of land holdings.
Crop Patterns: Types of crops grown and their rotation.
Economic Activities: Main sources of income for the community, which may influence irrigation needs.

5.4 Steps in Conducting the Survey:

Planning
Define the scope and objectives of the survey.
Assemble a team of experts (engineers, hydrologists, soil scientists, socio-economists).

Field Data Collection
Topographical Survey: Conduct detailed mapping using GPS and other surveying equipment.
Hydrological Survey: Install rain gauges, measure river flows, and monitor groundwater levels.
Soil Survey: Collect soil samples from different locations and depths for analysis.
Socio-Economic Survey: Conduct interviews and distribute questionnaires to gather data from the local community.

> ### Data Analysis
Topographical Data: Create detailed maps and models of the terrain.
Hydrological Data: Analyze water availability and variability.
Soil Data: Assess soil suitability for different types of crops and irrigation methods.
Socio-Economic Data: Understand community needs and potential impacts of the irrigation project.

Report Preparation
Compile all data and analysis into a comprehensive report.
Include maps, charts, and graphs to visually represent findings.
Provide recommendations for the design and implementation of the irrigation project.

5.5 Control level in reservoir:

Definition: Control level in a reservoir refers to specific water levels that are maintained to ensure efficient operation and safety.

Types of Control Levels:
Full Reservoir Level (FRL): The maximum water level that a reservoir can safely hold.
Minimum Draw down Level (MDL): The lowest water level at which water can still be supplied.
Dead Storage Level: The level below which water cannot be used because it is below the outlet pipes.

Purpose of Maintaining Control Levels:
Safety: Prevents overflow and potential flooding.
Operational Efficiency: Ensures there is enough water for supply, irrigation, and power generation.
Environmental Protection: Maintains a balance to protect aquatic ecosystems.

Factors Affecting Control Levels
Inflow: Amount of water entering the reservoir from rivers or rainfall.
Outflow: Amount of water being used or released from the reservoir.
Evaporation: Water loss due to heat.
Siltation: Sediment accumulation at the bottom of the reservoir.

Chapter 6: Dams & its Classification : Earthen & Gravity Dam

6.1 DAM:

6.1.1 Definition:
A dam is a barrier built across a river or stream to hold back water.
This creates a reservoir or lake, which can be used for various purposes like water supply, electricity generation, irrigation, and flood control.

6.1.2 Types of Dam:

Based on Material Used

Concrete Dams:
Gravity Dams: Made of concrete and rely on their weight to hold back water.
Arch Dams: Curved in shape, transferring water pressure to the sides of the canyon.

Earth Dams:
Made of compacted earth (soil and rock). They are heavy and rely on their weight to hold back water.

Rock-fill Dams:
Made of rocks and boulders, with an impervious core to prevent water from passing through.

Based on Purpose

Storage Dams:
Built to store water for irrigation, drinking water, industrial use, and recreation.

Diversion Dams:
Built to divert water from a river into a canal or irrigation system.

Detention Dams:
Built to control floodwater and reduce the risk of downstream flooding.

Hydro power Dams:
Built to generate electricity by using the stored water to turn turbines.

Based on Structure & Design

Fixed-Crest Dams:
Water flows over the top of the dam when the reservoir is full.

Movable Dams:
Have gates or other mechanisms to control the water flow.

Buttress Dams:
Have a sloping surface on the water side, supported by buttresses (supports) on the downstream side.

Based on Size

Large Dams:
Higher than 15 meters (about 49 feet) from the foundation to the crest.

Medium Dams:
Between 10 to 15 meters (about 33 to 49 feet) high.

Small Dams:
Less than 10 meters (about 33 feet) high.

Chapter 7: Earthen Dam: Components(Cross-Section)

7.1 CROSS-SECTION OF EARTHEN DAM:

A. Top Width
Definition: The horizontal width of the dam at the crest.
IS Specifications: IS 12169:1987 recommends a minimum top width of 5 meters for small dams and up to 10 meters or more for larger dams.
Design Considerations:
Accessibility for maintenance vehicles.
Safety and ease of operation during inspection.
Dependent Factors: Dam height, Functional requirements (e.g., roadways, emergency pathways).

B. Free Board
Definition: The vertical distance between the maximum water level and the top of the dam.
IS Specifications: IS 10635:1993 suggests a minimum free board of 1.5 to 3 meters, depending on wave action and potential for settlement.
Design Considerations:
Potential wave action due to wind.
Settlement and deformation of the dam body.
Additional height for safety during extreme weather conditions.
Dependent Factors:
Reservoir size and wind fetch.
Climatic conditions (e.g., rainfall, flooding potential).
Dam height and type.

Casing/Outer Shell
Definition: The outermost layer of the dam, typically composed of compacted earth.
IS Specifications: IS 8826:1978 provides guidelines for the selection of materials and compaction standards.
Design Considerations:
Selection of well-graded materials for stability.
Proper compaction to ensure durability and prevent erosion.
Dependent Factors:
Availability of construction materials.
Geo technical properties of the dam site.
Compaction methods and equipment.

Central Impervious Core
Definition: A central section of the dam made of impermeable material to prevent water seepage.
IS Specifications: IS 1498:1970 outlines the criteria for soil classification for impervious cores.
Design Considerations:
Material selection (clay or a mixture of clay and silt).
Thickness and placement to ensure maximum effectiveness.
Dependent Factors:
Height and hydraulic head of the dam.

Seepage potential and control requirements.
Quality and availability of impervious materials.

Cut-Off Trench
Definition: A trench filled with impervious material extending into the foundation to prevent seepage under the dam.
IS Specifications: IS 8414:1977 specifies the construction of cut-off trenches for earth and rock fill dams.
Design Considerations:
Depth and width to reach impermeable strata.
Ensuring continuity with the central core.
Dependent Factors:
Foundation geology.
Seepage characteristics and hydro static pressure.
Construction techniques and machinery.

Downstream Drainage System
Definition: A system of drains located on the downstream side to control and manage seepage.
IS Specifications: IS 9429:1999 provides guidelines for drainage systems in earth and rock fill dams.
Design Considerations:
Incorporation of horizontal drains, toe drains, and chimney drains.
Selection of permeable materials to facilitate drainage.
Dependent Factors:
Seepage rate and volume.
Geo technical conditions and permeability of dam materials.
Maintenance and inspection requirements.

Chapter 8: Gravity Dam:

8.1 GRAVITY DAM:

8.1.1 Definition:
A gravity dam is a massive structure made from concrete or stone masonry.
It holds back water by relying on its own weight to resist the horizontal thrust of the water pushing against it.

8.1.2 Forces Acting on Gravity Dam****:

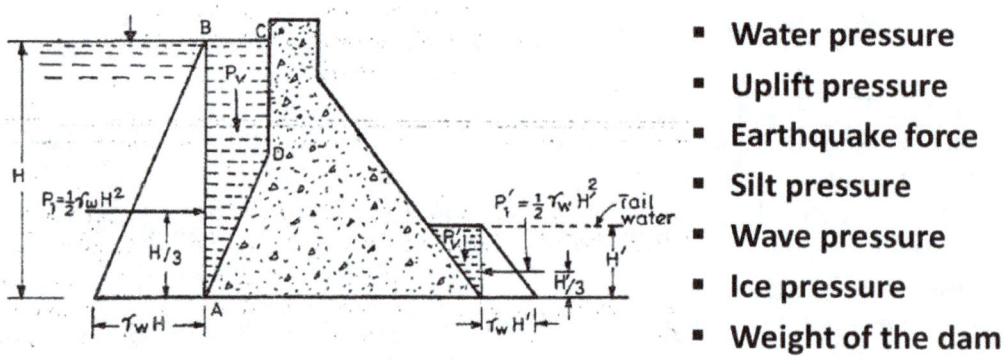

- Water pressure
- Uplift pressure
- Earthquake force
- Silt pressure
- Wave pressure
- Ice pressure
- Weight of the dam

Water Pressure (Hydro static Pressure)

Definition: The pressure exerted by the water on the dam.

Formulae: $P_w = \tfrac{1}{2}\gamma_w h^2$

Explanation:

- P_w = water pressure
- γ_w = unit weight of water (typically 9.81 kN/m³)
- h = height of water

Uplift Pressure:

Definition: The upward pressure exerted by water seeping underneath the dam.

Formula: $U = \gamma_w \times A \times h \times \frac{d}{l}$

Explanation:

- U = uplift pressure
- A = base area of the dam
- h = height of water column
- d = depth of dam foundation
- l = length of seepage path

Earthquake Forces:

Definition: Forces due to seismic activity, including both vertical and horizontal accelerations.

Horizontal Earthquake Force (Inertia Force):

- Formula: $E_h = k_h W$
- Explanation:
    - E_h = horizontal earthquake force
    - k_h = horizontal seismic coefficient
    - W = weight of the dam

Vertical Earthquake Force:

- Formula: $E_v = k_v W$
- Explanation:
    - E_v = vertical earthquake force
    - k_v = vertical seismic coefficient

Hydrodynamic Pressure:

Definition: Additional water pressure due to the motion of water during an earthquake.

Formula: $P_h = 0.875 \gamma_w h^2 k_h$

Explanation:

- P_h = hydrodynamic pressure
- γ_w = unit weight of water
- h = height of water
- k_h = horizontal seismic coefficient

Silt Pressure:

Definition: Pressure exerted by silt deposited against the dam.
- Formula: $P_s = \frac{1}{2} \gamma_s h_s^2$
- Explanation:
    - P_s = silt pressure
    - γ_s = unit weight of silt
    - h_s = height of silt

Wave Pressure:

Definition: Pressure due to wind-generated waves striking the dam.
- Formula: $P_w v = 2.4 \times 10^{-6} \sqrt{F} H_w^2$
- Explanation:
    - $P_w v$ = wave pressure
    - F = fetch length (distance over which the wind blows)
    - H_w = height of waves

Ice Pressure:
Definition: Force exerted by ice forming and expanding against the dam.
Typical Pressure: Can range from 250 to 1500 kN/m² depending on conditions.

Weight of Dam:

Definition: The force due to the mass of the dam itself.

Explanation:

- P_w = water pressure
- γ_w = unit weight of water (typically 9.81 kN/m³)
- h = height of water

8.1.3 Elementary Profile of Gravity Dam** :

Definition:
The elementary profile of a gravity dam is the simplest theoretical shape of the dam, usually a triangle, that allows us to understand and analyze the basic forces and stability requirements.

It is an idealized cross-section used to illustrate the fundamental concepts of dam design.

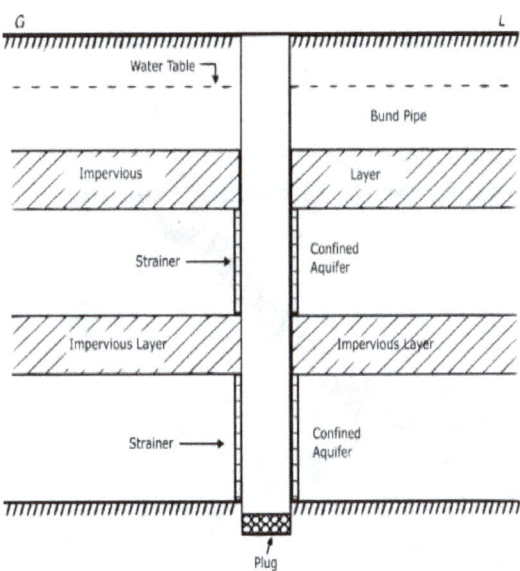

Forces acting on Elementary Profile:

Weight of Dam-
Definition: The downward force due to the dam's own weight.

- Formula: $W = \frac{1}{2}\gamma_c b h^2$
- Explanation:
  - γ_c = unit weight of concrete (typically 24 kN/m³)
  - b = base width of the dam
  - h = height of the dam

Water Pressure-
Definition: The horizontal pressure exerted by the water on the dam.

- Formula: $P_w = \frac{1}{2}\gamma_w h^2$
- Explanation:
  - γ_w = unit weight of water (typically 9.81 kN/m³)
  - h = height of water

Uplift Pressure-
Definition: The upward pressure exerted by water seeping underneath the dam.

- Formula: $U = \gamma_w b \frac{h}{2}$
- Explanation:
  - γ_w = unit weight of water
  - b = base width of the dam
  - h = height of water

Base Width of Elementary Profile:
To ensure stability, the base width of the elementary profile is determined using two criteria: stress criteria and stability criteria.

Stress Criteria-
Condition: The maximum compressive stress at the base of the dam should not exceed the allowable stress of the material.

- **Formula:** $\sigma_{max} = \frac{W}{b} + \frac{6M}{b^2}$
- **Explanation:**
  - σ_{max} = maximum compressive stress
  - W = weight of the dam
  - M = moment due to water pressure and uplift pressure about the toe

For no tensile stress at the heel:

- $b \geq \frac{h}{\sqrt{\frac{\sigma_c}{\gamma_c} - \frac{\gamma_w}{2\gamma_c}}}$
- Where σ_c is the allowable compressive stress.

Stability Criteria-
Condition: The dam should resist sliding and overturning.

- **Sliding Stability:**
  - Factor of safety against sliding: $F_s = \frac{\mu W}{P_w - U} \geq 1.5$
  - μ = coefficient of friction

- **Overturning Stability:**
  - Factor of safety against overturning: $F_o = \frac{W \times e}{M} \geq 2$
  - e = eccentricity of weight

Stress Developed in Elementary Profile:

- **Maximum Stress (σ_{max}):** At the toe (downstream edge of the base)
  - Formula: $\sigma_{max} = \frac{W}{b} + \frac{6M}{b^2}$
  - M is the moment due to water pressure and uplift pressure.

- **Minimum Stress (σ_{min}):** At the heel (upstream edge of the base)
  - Formula: $\sigma_{min} = \frac{W}{b} - \frac{6M}{b^2}$

Reservoir Empty Condition:

When the reservoir is empty, the forces acting on the dam are different:

**No water pressure on the upstream face.
** Reduced uplift pressure, possibly negligible.

Stress Analysis for Empty Reservoir:

**Only the weight of the dam acts downward.
** Maximum compressive stress at the toe due to dam's weight.

8.1.4 Practical Profile of Gravity Dam** :

Definition:
A practical profile of a gravity dam refers to the shape, dimensions, and structural characteristics that ensure the dam's stability and functionality under various conditions.

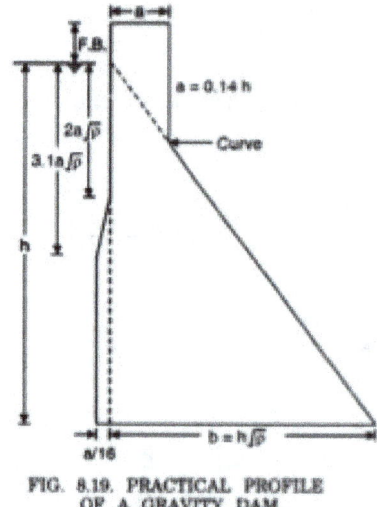

FIG. 8.19. PRACTICAL PROFILE OF A GRAVITY DAM.

Free board:
The margin given between Top level of Dam & High Flood Level in reservoir in order to ensure no wave splashing over the non overflow part.
It is kept as $3/2 h_w$.
The width of top section is kept 14% of height of dam.

IS Recommendations for Free board:
According to IS 6512-1972, the free board should be one and a half times the design wave height above the full reservoir level (FRL) or the maximum water level (MWL), whichever is higher, to ensure safety.
However, per the revised IS 6512-1984 recommendations, the free board should be equal to:

$$\tfrac{4}{3} h_w + \text{wind set-up}$$

where h_w is the wave height. The wave height and wind set-up should be calculated as per the method suggested by T. Saville.

Wind Setup Calculations:

Wind set-up is the piling up of water at one end of the reservoir due to wind action. It is determined by the following formula by Zuider Zee:

$$S = \frac{V^2 F \cos \beta}{62000 D}$$

where:

- S = wind set-up in meters
- V = wind velocity (km/hr)
- F = fetch (km)
- β = angle between wind direction and fetch
- D = average depth of water (m)

The wind velocity is typically taken as 120 km/hr for normal reservoir conditions and 80 km/hr for maximum water level conditions.

The free board should not be less than 1 meter above the maximum water level.

Modern practice suggests providing a maximum free board equal to 3 to 4% of the dam height, though a free board of 5% or more may be more economical.

8.1.5 High & Low Gravity Dam** :

Low Gravity Dam:

A low dam is designed so that the resultant of all forces passes through the middle third of its base, ensuring that the maximum compressive stress at the toe does not exceed permissible limits.

Limiting height (H) is given by:

$$H = \frac{f}{w(\rho - c + 1)}$$

where w is the unit weight of water, ρ is the density of dam material, c is the uplift factor, and f is the permissible stress.

For standard materials, the limiting height is approximately 88 meters.

High Gravity Dam:

A high dam exceeds the limiting height of a low dam.

To avoid excessive stresses, the dam's profile is adjusted by flattening the downstream slope and providing a batter on the upstream slope.

This adjustment ensures that the resultant forces remain near the center of the base, preventing excessive compressive stress and tension development.

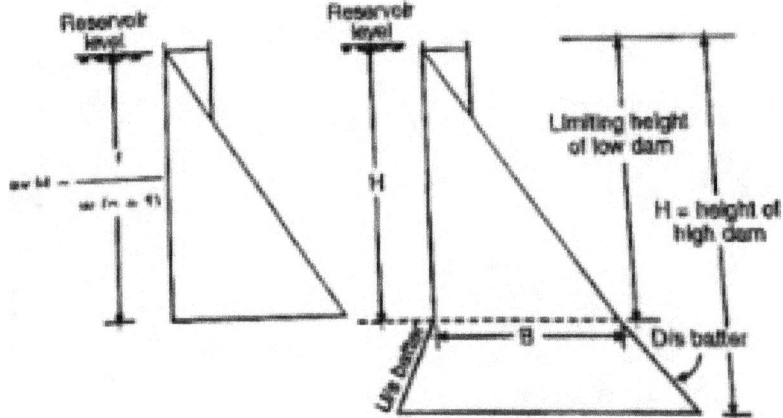

Chapter 9: Spillways

9.1 SPILLWAY:

Definition:
A spillway is a structure used to provide the controlled release of water from a dam or levee into a downstream area, typically a riverbed.
Its primary function is to prevent overflow of the dam and ensure safety.

Types of Spillway:

Ogee Spillway-
Shaped like an 'S' curve.
Allows smooth flow of water, reducing turbulence.
Commonly used in concrete dams.

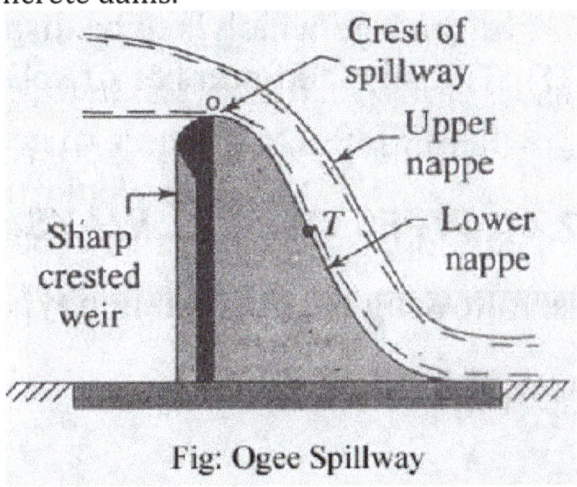

Fig: Ogee Spillway

Chute Spillway-
Also known as an open channel spillway.
Guides water along a channel to downstream areas.
Suitable for earth and rock fill dams.

Side Channel Spillway-
Water flows parallel to the dam before turning and dropping into a channel.
Used when there isn't enough space for a conventional spillway.

Shaft Spillway-
Water enters a vertical shaft and is then directed through a tunnel.
Effective for dams with limited space for spillways.

Siphon Spillway-
➤ Uses siphon action to automatically regulate the water flow.
➤ Begins to flow when water reaches a certain level.

1. **Discharge over an Ogee Spillway:**

$$Q = CLH^{3/2}$$

Where:

- Q = Discharge (cubic meters per second)
- C = Discharge coefficient
- L = Length of the spillway crest (meters)
- H = Head over the spillway crest (meters)

2. **Head Calculation:**

$$H = H_d + \frac{V^2}{2g}$$

Where:

- H_d = Depth of water above the spillway crest (meters)
- V = Velocity of approach (meters per second)
- g = Acceleration due to gravity (9.81 meters/second²)

3. **Shaft Spillway Capacity:**

$$Q = CA\sqrt{2gH}$$

Where:

- Q = Discharge (cubic meters per second)
- C = Coefficient of discharge
- A = Area of the shaft opening (square meters)
- H = Head (meters)

Chapter 10: Bandhara Irrigation

10.1 BANDHARA IRRIGATION:

10.1.1 Definition:
Bandhara irrigation is a traditional method of water management.
It is mainly used in regions with seasonal rivers or streams.
The system helps in storing and diverting water for agricultural use.
Employed in India, mainly in states of Maharashtra & Gujarat.

10.1.2 Layout & Components of Bandhara Irrigation:

CHECK DAM:
A small dam built across a stream or river to store and divert water.
Stores water during the rainy season, providing a steady supply for irrigation.
Can be made of stones, earth, concrete, or a combination.

SCOUR HOLE:
A deep pit formed downstream of the bandhara due to water flow.
Absorbs the energy of flowing water, reducing erosion.
Needs regular monitoring to ensure it does not become too deep or wide.

SCREEN WALL:
A wall built at the upstream side of the Bandhara.
Filters debris and sediments from the water before it enters the main canal.
Typically made of mesh or perforated material.

OUTLET:
A structure that allows controlled release of water from the Bandhara to the canals.
Regulates the amount of water flowing into the irrigation system.
Can include sluice gates or pipes.

FLOOD BANK:
Raised barriers on either side of the stream or river.
Prevents overflow and protects surrounding land from flooding.
Often made from earth or stone.

CANAL:
Main Canal:
The primary channel that carries water from the bandhara to the fields.
Distributes stored water to various parts of the irrigation system.
Dug out and lined to reduce water loss.

Field Channels:
Smaller channels branching from the main canal.
Directly deliver water to individual fields.
Must be carefully planned to ensure even distribution.

LINING WALL:
A protective wall lining the canals.
Prevents seepage and erosion, ensuring efficient water flow.
Can be made from concrete, bricks, or stones.

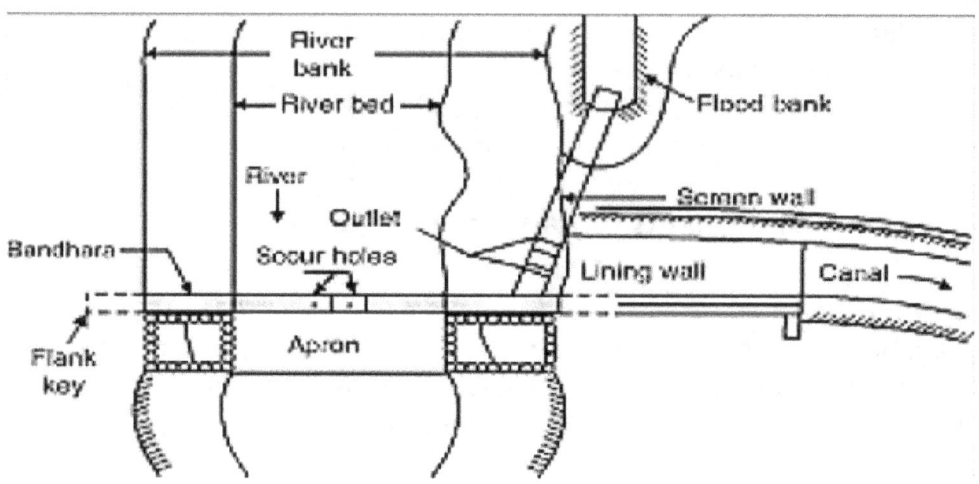

10.1.3 Construction of Bandhara Irrigation:
Site Selection:
Choose a location with a seasonal stream and suitable topography.

Building the Bandhara:
Construct a small dam across the stream.
Use materials like stones, earth, or concrete.

Digging Canals:
Create a main canal from the bandhara to the fields.
Dig smaller field channels branching off the main canal.

Constructing Spillways:
Build spillways to manage overflow water.

10.1.4 Working Mechanism of Bandhara Irrigation:
1. Water Collection:
During the rainy season, water is stored behind the bandhara.

Water Distribution:
Water is released from the Bandhara into the main canal.
The main canal carries water to field channels.
Field channels distribute water to the agricultural lands.

Managing Water Flow:
Spillways prevent overflow by diverting excess water.
Embankments ensure water flows in the desired direction.

Chapter 11: Percolation Tank

11.1 PERCOLATION TANK:

11.1.1 Definition:
A percolation tank is a shallow, artificial pond that helps store rainwater.
The stored water gradually seeps (percolates) into the ground, replenishing the groundwater levels.

11.1.2 Need of Percolation Tank:

1. Groundwater Recharge:
Helps to increase the amount of water stored underground.
Important in areas where groundwater levels are falling.

Drought Prevention:
Provides a source of water during dry periods.
Helps maintain water availability for agriculture and drinking.

Flood Control:
Reduces the risk of floods by capturing excess rainwater.
Slows down water flow, minimizing soil erosion.

Sustainable Water Supply:
Ensures a continuous supply of water throughout the year.
Supports ecosystems and human activities.

11.1.3 Selection of Site for Percolation Tank:

Topography:
Choose a gently sloping area to ensure easy water collection.
Avoid steep slopes where water would flow too quickly.

Soil Type:
Ideal soil should be sandy or loamy, allowing water to seep through easily.
Avoid rocky or clayey soils that don't allow water to percolate.

Rainfall:
Select areas that receive moderate to high rainfall to fill the tank.
Ensure that the site can capture runoff from nearby areas.

Water Table:
The water table should be low enough to allow storage but not too deep.
This ensures the stored water effectively recharges the groundwater.

Proximity to Water Sources:
Choose a site close to natural water sources like streams or rivers.
Helps to capture and store more water during rains.

Land Ownership:
Ensure the land is publicly owned or that there is permission from private owners.
This avoids legal issues and ensures community access.

Environmental Impact:
Assess the impact on local flora and fauna.
Ensure that the tank does not disrupt the natural habitat excessively.

Chapter 12: Lift Irrigation Scheme: Layout, Components, Function

12.1 LIFT IRRIGATION SCHEME:

12.1.1 Definition:

Lift irrigation is a method of watering crops by lifting water from a lower level (like a river or well) to a higher level.

It uses pumps and other equipment to move water to fields that are at a higher elevation.

12.1.2 Layout & Components of Lift Irrigation Scheme:

Water Source:
Usually a river, lake, reservoir, or well where water is available.
This is the starting point for the irrigation system.

Pumping Station:
A place near the water source where pumps are installed to lift water.
Includes a power supply to operate the pumps.

Rising Main:
A pipeline that carries water from the pumping station to a higher elevation.
This is typically a strong, durable pipe to withstand the pressure.

Distribution Network:
A series of smaller pipes or channels that distribute water to the fields.
Can include canals, ditches, or drip irrigation systems.

Field Application:
The final step where water reaches the crops.
Methods include sprinklers, drip systems, or simple flooding of fields.

12.1.3 Functions of Lift Irrigation Scheme:

Water Supply:
Provides a reliable water source for irrigation, especially in areas without natural gravity flow.
Ensures crops get enough water even during dry periods.

Increased Agricultural Productivity:
Enables cultivation of land that would otherwise be dry and unproductive.
Leads to higher crop yields and more stable food production.

Efficient Water Use:
Allows precise control of water distribution.
Minimizes water wastage by delivering water directly to where it's needed.

Flexibility in Irrigation:

Can be used to irrigate fields at different elevations and distances from the water source.
Adaptable to various types of crops and farming practices.

Support for Multiple Cropping:
Provides water throughout the year, enabling farmers to grow multiple crops.
Increases overall farm income and food security.

Chapter 13 : Drift & Sprinkler Irrigation

13.1 DRIP IRRIGATION SCHEME:

13.1.1 Definition:
Drip irrigation delivers water directly to the base of plants through small tubes and emitters.
It conserves water by minimizing evaporation and runoff.

13.1.2 Layout & Components of Drip Irrigation Scheme:

Water Source:
A well, reservoir, or other reliable water source.

Pump:
Moves water from the source through the system.
Ensures steady water pressure.

Mainline Pipe:
A large pipe that carries water from the pump to the field.
Connects to smaller pipes (lateral lines).

Lateral Lines:
Smaller pipes that run along rows of crops.
Distribute water to individual plants.

Emitters/Drippers:
Small devices attached to lateral lines.
Release water slowly and directly to the roots of plants.

Control Valves:
Regulate the flow of water through the system.
Allow different sections to be watered separately.

Filters:
Clean the water before it enters the system.
Prevent clogging of emitters.

13.1.3 Necessity of Drip Irrigation Scheme:

Water Conservation:
Uses less water by targeting the root zone.
Reduces evaporation and runoff.

Improved Plant Growth:
Provides consistent moisture.
Reduces plant stress and increases yields.

Reduced Weeds:
Limits water to the plant area only.
Reduces weed growth between rows.

Nutrient Management:
Allows for precise application of fertilizers.
Improves nutrient uptake by plants.

13.2 SPRINKLER IRRIGATION SCHEME:

13.2.1 Definition:
Sprinkler irrigation sprays water over the plants like natural rainfall using overhead sprinklers.

13.2.2 Layout & Components of Sprinkler Irrigation:

Water Source:
A well, reservoir, or other reliable water source.

Pump:
Moves water from the source through the system.
Ensures steady water pressure.

Mainline Pipe:
A large pipe that carries water from the pump to the field.
Connects to smaller pipes (lateral lines).

Lateral Lines:
Smaller pipes that distribute water to sprinklers.
Often placed along the edges or through the middle of the fields.

Sprinklers:
Devices that spray water over the crops.
Can be fixed, rotating, or moving types.

Control Valves:
Regulate the flow of water through the system.
Allow different sections to be watered separately.

13.2.3 Necessity of Drip & Sprinkler Irrigation:

Efficient Water Use:
Both systems save water by reducing evaporation and runoff.
Important in areas with limited water resources.

Improved Crop Yields:
Provide consistent and precise watering.
Enhance plant growth and productivity.

Cost Savings:
Reduce water and labor costs.
Lower the need for fertilizers and pesticides through targeted application.

Environmental Benefits:
Minimize soil erosion and nutrient runoff.
Promote sustainable farming practices.

Flexibility and Adaptability:
Can be used for various crops and field conditions.
Easily adjusted to changing water needs and weather conditions.

Chapter 14 : Well Irrigation

14.1 WELL IRRIGATION:

14.1.1 Definition:
Well irrigation involves using water from wells to irrigate crops.
Wells are deep holes drilled into the ground to access underground water.

14.1.2 Types of Well Irrigation:

Dug Wells:
Shallow wells dug by hand or using simple machinery.
Typically not very deep (10-20 meters).
Relies on groundwater close to the surface.

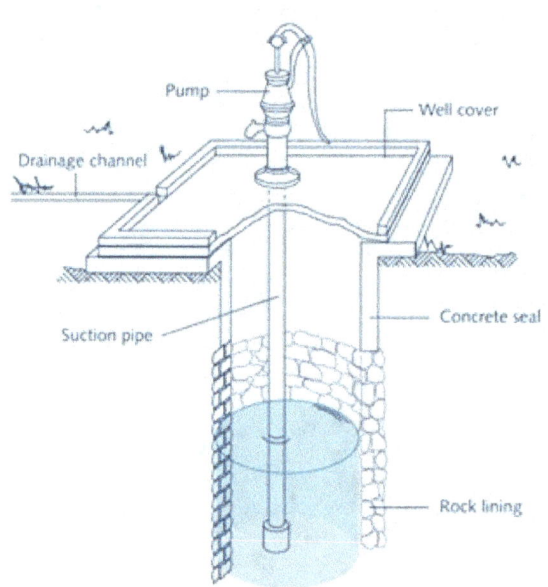

Tube Wells:
Deep wells created using drilling equipment.
Can reach depths of over 100 meters.
Uses a pump to bring water to the surface.
More reliable and can access deeper groundwater.
It is of following types:
Slotted Tube Well
Cavity Tube Well
Strainer Tube Well

Slotted Tube Well:
A slotted tube well has slots or narrow openings cut into the tube, allowing water to enter from the surrounding aquifer.
Components:
Casing Pipe:
A solid pipe installed in the drilled hole.

Supports the well structure and prevents collapse.

Slotted Section:
The bottom section of the tube has narrow slots or openings.
Allows water to enter the well from the aquifer.

Pump:
Installed at the top or within the well to draw water up.

Cavity Tube Well:
A cavity tube well creates a cavity or open space at the bottom of the well to allow water to collect and enter the tube.
Components:
Casing Pipe:
A solid pipe installed in the drilled hole.
Supports the well structure.

Cavity:
An open space or widened section at the bottom of the well.
Allows water to collect before entering the tube.

Pump:
Installed at the top or within the well to lift water.

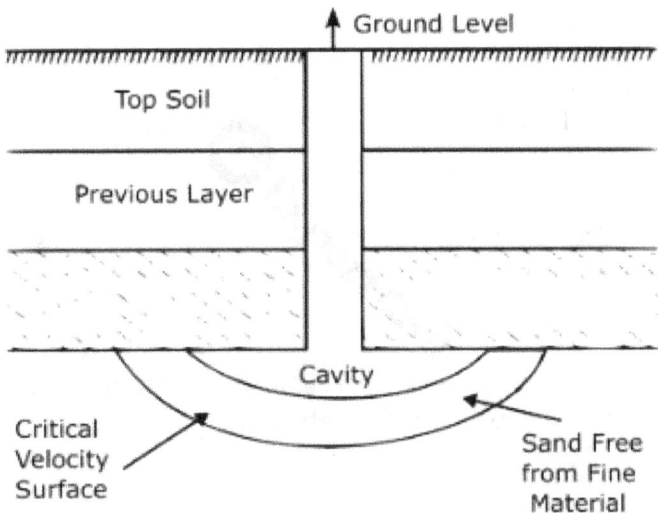

Strainer Tube Well:
A strainer tube well uses a strainer (a perforated section of the tube) to filter out sand and other particles from the water.
Components:
Casing Pipe:
A solid pipe installed in the drilled hole.
Prevents the walls of the hole from collapsing.

Strainer:
A perforated or slotted section of the tube near the bottom.

Allows water to enter while keeping out sand and debris.

Pump:
Installed at the top or within the well to lift water to the surface.

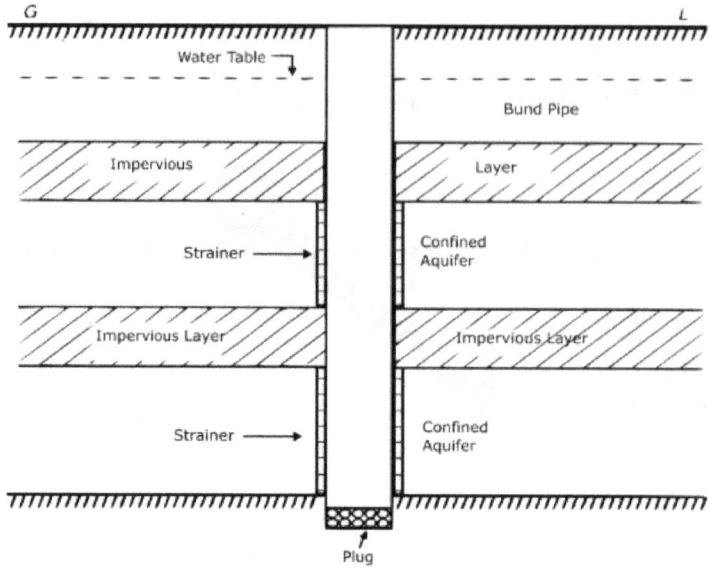

14.1.3 Advantages & Disadvantages of Well Irrigation:

Advantages of Well Irrigation:

Reliable Water Source:
Provides a consistent supply of water.
Less affected by seasonal changes and droughts compared to surface water.

Accessibility:
Can be built close to fields, reducing the need for long-distance water transportation.
Provides water on-demand, whenever needed.

Cost-Effective:
Lower initial cost compared to large-scale irrigation systems like canals.
Maintenance and operation costs are relatively low.

Improved Crop Yields:
Allows for controlled and efficient water use.
Enhances crop growth and productivity.

Flexibility:
Suitable for small and large farms.
Can be used to irrigate various types of crops.

Disadvantages of Well Irrigation:

Depletion of Groundwater:
Excessive use can lower the water table.
Can lead to dry wells and reduced water availability over time.

High Initial Investment for Tube Wells:
Drilling deep wells and installing pumps can be expensive.
Requires significant upfront capital.

Salinity and Contamination:
In some areas, groundwater may be saline or contaminated.
Can affect soil health and crop quality.

Energy Costs:
Pumping water requires energy, usually electricity or diesel.
Increases operational costs.

Maintenance:
Wells and pumps require regular maintenance.
Costs and effort needed to keep the system operational.

14.1.4 Yield of Well :

4.6.4.1 Definition:
The yield of a well refers to the amount of water it can supply.
Measured in liters per second (L/s) or gallons per minute (GPM).

4.6.4.1 Factors affecting Yield:

Aquifer Properties:
The type of rock or soil where the well is drilled.
Permeability and porosity affect water flow.

Well Depth:
Deeper wells can access more water.
Generally, deeper wells have a higher yield.

Pump Efficiency:
The efficiency of the pump used to draw water.
Better pumps can extract water more effectively.

Recharge Rate:
The rate at which the aquifer replenishes.
Affects the sustainable yield of the well.

4.6.4.3 Formula for Well Yield:

Specific Capacity = Yield (Q) / Draw down (s)

Where:
Q = Yield of the well (volume/time, e.g., L/s or GPM)
s = Draw down (the drop in water level in the well during pumping, measured in meters or feet)

Chapter 15: Weirs

15.1 WEIRS:

15.1.1 Definition:
A weir is a barrier built across a river or stream to control the flow of water. It helps measure the flow rate, raise water levels, or divert water.

15.1.2 Types of Weirs

1. Depending on the Opening Shape

Rectangular Weir:
The opening has a rectangular shape.
It is easy to construct and commonly used for flow measurement.

Triangular Weir:
The opening has a triangular shape.
Often called a V-notch weir.
Useful for measuring small flow rates with high accuracy.

Trapezoidal Weir:
The opening has a trapezoidal shape.
Known as a Cipoletti weir.
Designed to provide a more accurate flow rate by reducing errors due to the contraction of water flow at the sides.

2. Depending on Crest Form

Sharp-Crested Weir:
The crest (top edge) is very thin and sharp.
Water flows over it smoothly, making it useful for precise flow measurements.

Broad-Crested Weir:
The crest is wide and flat.
Water flows over it with a more gradual transition.
Suitable for high-flow conditions

Narrow-Crested Weir:
The crest is narrow, but not as sharp as a sharp-crested weir.
Balances the benefits of both sharp and broad-crested weirs.

Ogee-Shaped Weir:
The crest has a curved shape (like an "S" turned on its side).
Designed to reduce energy loss and provide smooth flow over the crest.
Often used in spillways and dam structures.

3. Depending on the Effects of Side on Nappe

Weir with End Contraction:
The sides of the weir opening are not connected to the channel walls.
Water contracts horizontally as it flows through the weir.

Useful in situations where side contractions are expected and need to be measured.

Weir without End Contraction:
The sides of the weir opening are connected directly to the channel walls.
There is no horizontal contraction of water.
Used when a more uniform and straightforward flow measurement is required.

15.1.3 Components of Weir:
Weir has following components:

Crest
Definition: The top edge or surface of the weir over which water flows.
Function: Determines the height of the water that flows over the weir, which helps in measuring the flow rate.

Shutter
Definition: A movable barrier used to control the water flow over the crest.
Function: Adjusts the height of the water and flow rate as needed.

Weir Wall
Definition: The vertical structure that supports the crest and shutters.
Function: Acts as the main barrier to control water flow and maintain the structure's stability.

Launching Apron
Definition: A layer of stones or concrete placed on the downstream side of the weir.
Function: Prevents erosion by dissipating the energy of falling water as it exits the weir.

Impervious Apron
Definition: A waterproof layer, usually made of concrete or clay, laid on the upstream and downstream sides.
Function: Prevents water from seeping under the weir, which could cause undermining and instability.

F. Inverted Filter
Definition: A layer of graded gravel or coarse sand placed beneath the impervious apron.
Function: Allows water to drain away while preventing soil particles from being washed out, thus maintaining the weir's stability.

Upstream (U/S) Pile
Definition: Piles driven into the riverbed on the upstream side of the weir.
Function: Supports the weir structure and helps prevent water from seeping underneath.

Downstream (D/S) Pile
Definition: Piles driven into the riverbed on the downstream side of the weir.
Function: Provides additional support and helps secure the launching apron in place.

Block Protection
Definition: Blocks of concrete or stone placed around the weir.
Function: Protects the weir from damage caused by debris and water flow, and helps prevent erosion.

Chapter 16: Barrages

16.1 BARRAGE:

16.1.1 Definition:
A barrage is a barrier constructed across a river to control and manage the flow and level of water.
It is essential for irrigation, water supply, flood control, and navigation.

16.1.2 Components of Barrage:

Main Barrage Portion:
The primary structure that spans the river.
Made of concrete or masonry.
Includes gates and spillways to control water flow.

Divide Wall:
A wall that separates different sections of the barrage.
Helps to reduce the pressure on the main structure.
Typically made of reinforced concrete.

Fish Ladder:
Series of steps or pools allowing fish to bypass the barrage.
Essential for the migration of fish.
Made of concrete, with gentle slopes to facilitate fish movement.

Sheet Piles:
Vertical walls driven into the riverbed.
Prevent seepage of water through the foundation.
Made of steel or reinforced concrete.

Upstream Piles:
Piles placed upstream to support the barrage.
Help in stabilizing the structure against water pressure.
Typically made of steel or concrete.

Intermediate Sheet Piles:
Piles placed between upstream and downstream to add stability.
Prevent seepage and increase the barrage's strength.
Made of steel or reinforced concrete.

Inverted Filter:
Layer of sand and gravel under the barrage.
Prevents the movement of fine soil particles, reducing seepage.
Ensures stability of the structure.

Flexible Apron:

Layer of concrete or stones downstream of the barrage.
Absorbs energy of falling water to prevent erosion.
Often extends several meters downstream.

Under Sluices:
Gates located at the bottom of the barrage.
Used to flush out silt and control sediment build-up.
Made of steel, can be raised or lowered.

Guide Banks:
Structures parallel to the river, guiding water towards the barrage.
Prevents erosion and directs flow smoothly.
Made of earth or stones, often with protective cover.

Marginal Bunds:
Embankments built along the riverbanks.
Protect adjacent land from flooding.
Made of compacted earth, sometimes reinforced with stones.

Groynes or Spurs:
Structures projecting into the river from the banks.
Direct the flow away from vulnerable areas, reducing erosion.
Made of stones or concrete.

*Specifications

Length of Barrage: Can vary from 100 meters to several kilometers.
Height of Barrage: Typically ranges from 5 to 20 meters.
Materials: Commonly concrete, steel, and masonry.
Gate Size: Varies, often between 2 to 5 meters wide and 4 to 6 meters high.
Fish Ladder Slope: Usually 1:10 to 1:20 to allow easy fish movement.
Sheet Pile Depth: Can be up to 20 meters deep depending on soil conditions.
Flexible Apron Length: Often extends 20 to 50 meters downstream.

16.1.3 DIFFERENCE BETWEEN WEIR & BARRAGE:

Feature	Weir	Barrage
Definition	A low wall built across a river to raise water levels.	A barrier with adjustable gates built across a river to control water flow and levels.
Primary Purpose	Mainly to raise the water level.	To control and regulate water flow for multiple purposes such as irrigation, flood control, and navigation.
Structure	Fixed structure with no moving parts.	Includes adjustable gates and sometimes movable parts.
Water Flow Control	Limited control, primarily allows water to overflow.	High control over water flow due to adjustable gates.
Complexity	Simpler and less complex.	More complex with sophisticated gate mechanisms.
Construction Cost	Generally lower cost.	Higher cost due to complexity and materials.
Flexibility	Low flexibility; fixed water level adjustment.	High flexibility; can adjust water levels and flow as needed.
Common Uses	Raising water levels, simple diversion for small-scale irrigation.	Irrigation, flood control, navigation, water supply, and sometimes power generation.
Height	Usually lower, often just a few meters high.	Can be higher, often designed to manage significant water levels.
Environmental Impact	Lesser impact on fish migration and sediment flow.	Can have a greater impact due to significant alteration of river flow, often includes fish ladders to mitigate effects.
Examples	Used in small rivers or streams, traditional designs.	Used in larger rivers, modern and multi-functional designs.

Chapter 17: Diversion Head works

17.1 DIVERSION HEADWORKS:

17.1.1 Definition:
Diversion head works are structures built across a river or stream to divert water into a canal or other water distribution system for purposes such as irrigation, water supply, or power generation.

These structures include components that control and manage the water flow, ensuring efficient and sustainable use of the water resource.

17.1.2 Components of Diversion Head works:

1. Weir/Barrage
Function: Controls the flow and level of the river.
Details:
*A barrier built across the river.
*Raises the water level upstream to divert water into a canal.
*Can be fixed (weir) or adjustable (barrage with gates).

2. Divide Wall
Function: Separates different parts of the water flow to reduce pressure on the main structure.
Details:
*A wall extending from the weir/barrage downstream.
*Directs water flow towards specific areas like sluices and canals.
*Helps in sediment control and flow management.

3. Fish Ladder
Function: Allows fish to bypass the weir/barrage and continue their natural migration.
Details:
*A series of steps or small pools.
*Provides a gentle slope for fish to move upstream.
*Important for maintaining aquatic life.

4. Approach Channel
Function: Directs water smoothly towards the weir/barrage and other components.
Details:
*A channel leading water from the river to the weir/barrage.
*Ensures a steady and controlled approach flow.
*Helps in reducing turbulence and sediment deposition.

5. Scouring Sluices
Function: Remove sediment and silt from the water near the weir/barrage.
Details:
*Gates or openings near the bottom of the structure.
*Allow controlled release of water and sediment.
*Helps keep the upstream side clear of silt.

6. Silt Prevention Devices
Function: Prevent excessive silt from entering the canal system.

Details:
*Structures like silt excluders or silt ejectors.
*Direct silt-laden water away from the canal intake.
*Keeps the water clean for irrigation and other uses.

7. Canal Head Regulator
Function: Controls the flow of water from the river into the canal.
Details:
*Gates located at the head of the canal.
*Regulates the amount of water entering the canal.
*Ensures a consistent water supply for irrigation.

8. Marginal Bunds
Function: Protect the surrounding land from flooding.
Details:
*Earthen embankments built along the sides of the river.
*Prevent floodwater from spilling into adjacent areas.
*Provide additional stability and protection to the main structure.

Chapter 18: Canals

18.1 CANAL:

18.1.1 Definition:
A canal is a man-made waterway constructed to carry water for various purposes such as irrigation, transportation, drainage, or water supply.
Canals help distribute water from rivers or reservoirs to agricultural fields, urban areas, or other destinations.

18.1.2 Classification of Canals:

A. According to Alignment
1. Contour Canal
Definition: A canal that follows the natural contours or elevation levels of the land.
Characteristics:
*Runs along the contour lines.
*Maintains a consistent elevation.
*Minimizes excavation and construction costs.
Use: Commonly used for irrigation as it efficiently follows the terrain.

Ridge Canal
Definition: A canal constructed along the ridge or highest points of the terrain.
Characteristics:
*Positioned at the top of ridges.
*Water flows by gravity to either side of the ridge.
*Requires lifting water to the ridge, often using pumps.
Use: Used to supply water to areas on both sides of the ridge.

Side Slope Canal
Definition: A canal built along the slopes of hills or elevated areas.
Characteristics:
*Runs parallel to the slope of the land.
*Utilizes gravity for water flow.
*Requires careful engineering to prevent erosion and maintain stability.
Use: Suitable for hilly or undulating terrains.

According to Position in Canal Network
Main Canal
Definition: The primary canal that carries water directly from the source (e.g., a river, reservoir).
Characteristics:
*Largest canal in the network.
*Delivers water to branch canals.
*High capacity and significant length.
Use: Main artery of the canal system, distributing water to various regions.

Branch Canal
Definition: A canal that branches off from the main canal to supply water to smaller areas.
Characteristics:
*Intermediate size between main and distributary canals.
*Feeds water to distributary canals.

*Spreads water over a wider area.
Use: Extends the reach of the main canal, serving secondary regions.

Distributary Canal
Definition: A smaller canal that takes water from branch canals to specific fields or areas.
Characteristics:
*Smallest canals in the network.
*Directly supply water to the agricultural fields.
*Shorter in length, with lower capacity.
Use: Final delivery of water to the end users, usually farmers.

Minor Canal
Definition: A canal that further distributes water from distributary canals to individual plots.
Characteristics:
*Very small in size.
*Serve small areas or individual fields.
*Low capacity, often with manual control structures.
Use: Ensures water reaches every part of the irrigation system.

Field Channel
Definition: Small channels that directly distribute water from minor canals to the fields.
Characteristics:
*Smallest and simplest in the network.
*Usually unlined and temporary.
*Managed by farmers for direct irrigation.
Use: Direct water to crops, providing the final stage of water distribution.

18.2 DESIGN OF MOST ECONOMICAL CANAL SECTION:

Designing the most economical canal section involves ensuring that the canal can carry the required discharge with the minimum cross-sectional area to reduce construction costs.
This procedure typically applies to a trapezoidal canal section, which is the most common in practice.

Step-by-Step Numerical Procedure
Determine the Design Discharge (Q)
Calculate or obtain the required discharge, ☐ (in cubic meters per second, m^3/s).

Select the Channel Slope (S)
Choose an appropriate channel bed slope, S (dimensionless), based on the topography and design standards.

Choose the Manning's Roughness Coefficient (n)

Select the Manning's roughness coefficient, n, which depends on the type of canal lining (dimensionless).

Calculate the Hydraulic Radius (R)
Use the formula for the hydraulic radius, R:

$$R = \frac{A}{P}$$

where A is the cross-sectional area (m²) and
P is the wetted perimeter (m).

Use Manning's Equation
Apply Manning's equation to relate discharge, hydraulic radius, slope, and roughness coefficient:

$$Q = \frac{1}{n} A R^{2/3} S^{1/2}$$

Express Area and Perimeter in Terms of Variables

For a trapezoidal section, express the area A and wetted perimeter P in terms of bottom width b, depth y, and side slope z (horizontal to vertical):

$$A = by + zy^2$$

$$P = b + 2y\sqrt{1+z^2}$$

Optimize the Section for Minimum Wetted Perimeter
For a trapezoidal canal, the most economical section occurs when the hydraulic radius R is maximized, which happens when the wetted perimeter minimized for a given area A
The condition for the most economical section for a trapezoidal channel is:

$$b = 2y\sqrt{1+z^2}$$

Substitute and Solve for Depth (y)

$$A = 2y^2\sqrt{1+z^2}$$

$$Q = \frac{1}{n}(2y^2\sqrt{1+z^2}) \left(\frac{2y\sqrt{1+z^2}}{4y\sqrt{1+z^2}}\right)^{2/3} S^{1/2}$$

Simplify and Solve for Bottom Width (b)

$$b = 2y\sqrt{1 + z^2}$$

Verify Design and Adjust If Necessary
Check the calculated dimensions ☐ and ☐ to ensure they meet practical design constraints and standards.

Adjust if necessary for considerations such as maintenance, safety, and local conditions.

Calculate Final Parameters

$$A = by + zy^2$$

$$P = b + 2y\sqrt{1 + z^2}$$

Check Flow Velocity and Stability

$$V = \frac{Q}{A}$$

Cross drainage work-aqueduct, syphon aqueduct, super passage, level crossing

18.3 CROSS-DRAINAGE WORKS:

Cross-drainage works are structures designed to facilitate the safe and efficient passage of water from natural or artificial channels (such as streams, rivers, or irrigation canals) across transportation routes (like roads, railways, or canals).

These works ensure that the intersecting infrastructure and water flow systems do not negatively impact each other.

18.4 AQUEDUCT:

In case of Aqueduct, H.F.L(High Flood Level) of the drain is much below bottom of canal so that drainage water flows freely under gravity.

An aqueduct consists of a masonry or concrete trough of rectangular section supported on abutments of piers.

The drain water flows below the trough through the abutments and piers. In this type, the canal is open for section.

A trough is divided into compartments and roadway is provided.

The width of trough is made less than the width of canal to reduce the cost of masonry structure.

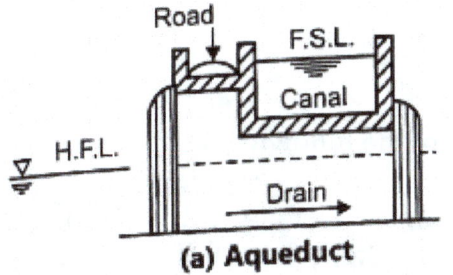

(a) Aqueduct

18.5 SYPHON AQUEDUCT:

When the high flood level of the drain is much higher above the canal the usual type of aqueduct cannot be provided, it is called Syphon Aqueduct.

The bed of the drain expressed below the crossing to form internal siphon. The drain water flows under hydro-static pressure.

Necessary slope given to the depressed drain to give self-cleaning velocity so as to avoid slitting.

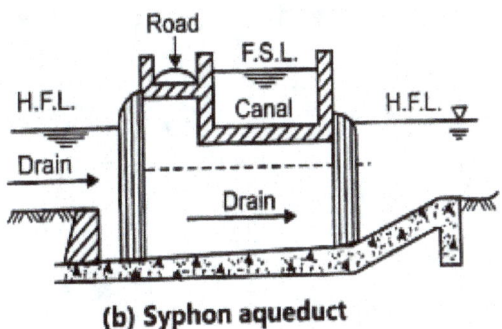

(b) Syphon aqueduct

18.6 SUPER PASSAGE:

The hydraulic structure in which the drainage Is passing over the irrigation canal is known as super passage.

This structure is suitable when the bed level of drainage is above the flood surface level of the canal.

The water of the canal passes clearly below the drainage.

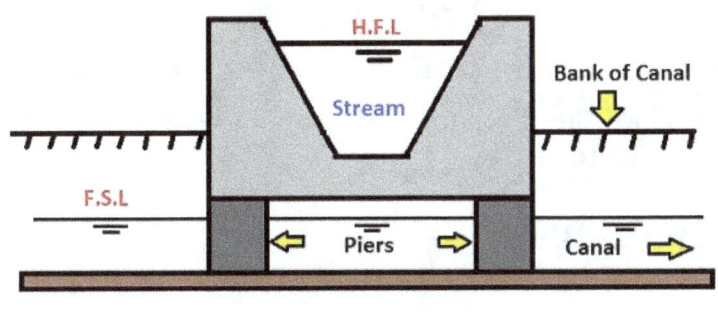

Supper Passage

18.7 LEVEL CROSSING:

The level crossing is an arrangement provided to regulate the flow of water through the drainage and the canal when they cross each other approximately at the same bed level.

The level crossing consists of the following components.

Crest wall

It is provided across the drainage just at the upstream side of the crossing point. The top level of the crest wall is kept at the full supply level of the canal.

Drainage regulator

It is provided across the drainage just at the downstream side of the crossing point. The regulator consists of adjustable shutters at different tiers.

Canal regulator

It is provided across the canal just at the downstream side of the crossing point. This regulator also consists of adjustable shutters at different tiers.

Inlet and outlet

In the crossing of small drainage with a small channel, Simple openings are provided for the flow of water in their respective directions. This arrangement is known as inlet and outlet.

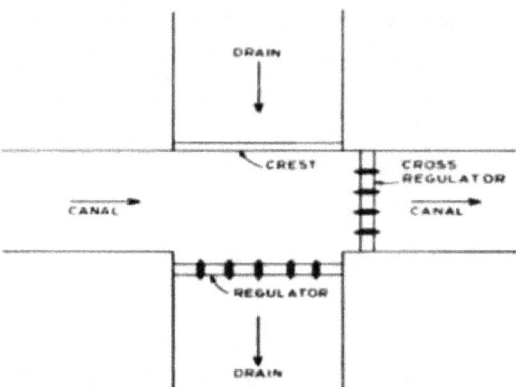

Chapter 19: Canal regulator- head regulator, cross regulator, escape, falls and outlets

19.1 CANAL REGULATOR:

Any structure to regulate the discharge, full supply level or velocity in a canal is known as Canal Regulator.

TYPES OF CANAL REGULATOR:

CANAL FALL-

A canal fall is a structure built on a canal to lower the water level from one level to another.

It helps in controlling the flow and speed of water, preventing erosion and damage to the canal.

Types of Canal Fall

Ogee Fall:
Shape: Curved like an 'S' or an 'ogee' shape.
Function: Smoothens the transition of water from higher to lower levels.
Usage: Commonly used due to its efficiency in reducing erosion.

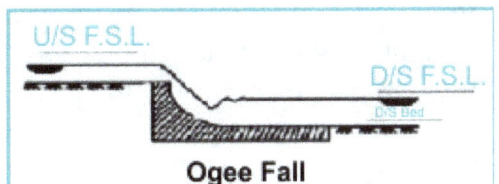

Rapid Fall:
Structure: Steep slope.
Water Flow: Water flows down quickly, resembling a waterfall.
Usage: Used where a quick drop in water level is needed.

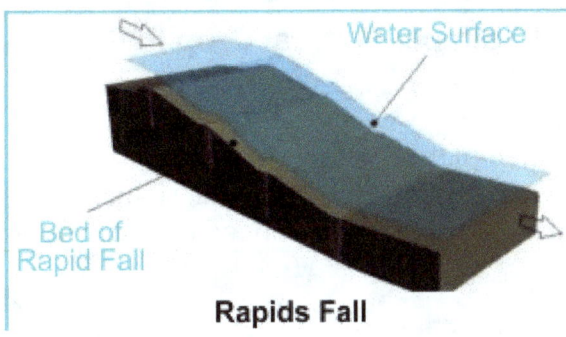

Stepped Fall:
Design: Series of steps or terraces.
Water Flow: Water flows down step by step, reducing speed gradually.
Benefit: Controls erosion better by breaking the energy of flowing water at each step.

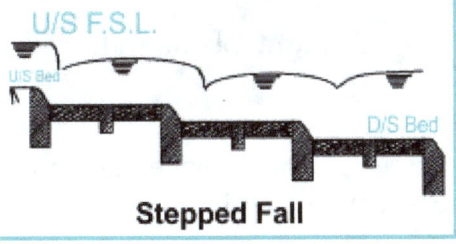

Notch Fall:
Structure: Notches or cut-outs in the fall wall.
Water Flow: Water flows through these notches, distributing flow and reducing force.
Application: Suitable for smaller canals with less water flow.

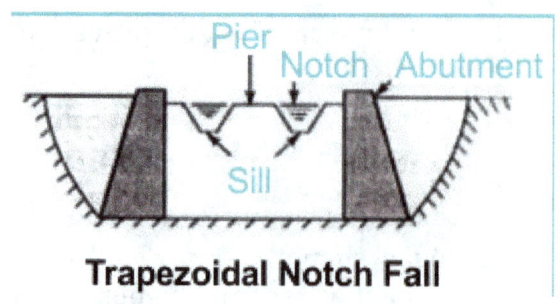

Trapezoidal Notch Fall

Vertical Drop Fall:

Design: Vertical wall or drop.
Water Flow: Water drops straight down from a height.
Consideration: Effective but can cause erosion at the base, needing protection measures.

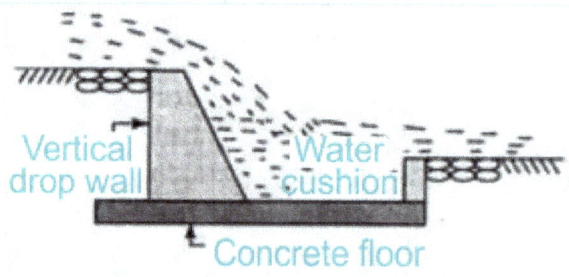

Simple Vertical Drop Fall

Glacis Type Fall:
Structure: Sloped or inclined plane.
Water Flow: Smooth flow down the inclined plane, reducing speed gradually.
Benefit: Minimizes erosion by reducing the speed of water flow smoothly.

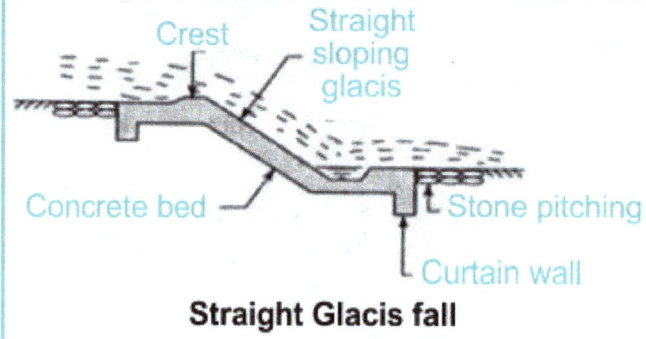

Straight Glacis fall

Meter Fall:
Definition: Falls equipped with measuring devices to monitor water flow.
Usage: Used for precise measurement and control of water discharge in irrigation systems.

Non-Meter Fall:
Definition: Falls without any measuring devices.
Usage: Used in areas where precise water measurement is not critical.

Advantages of Canal Fall

Controls water flow and minimizes erosion.

Facilitates energy dissipation.

Enables water level regulation.

Disadvantages of Canal Fall

Construction cost.

Maintenance and repair expenses.

Potential for sediment accumulation.

19.2 CANAL ESCAPE:

A canal escape is a specifically designed system or channel within a canal network that enables the safe diversion or release of excess water.

This feature is essential for preventing flooding and maintaining the canal's stability during periods of high flow, such as heavy rainfall or other conditions that could lead to overflow.

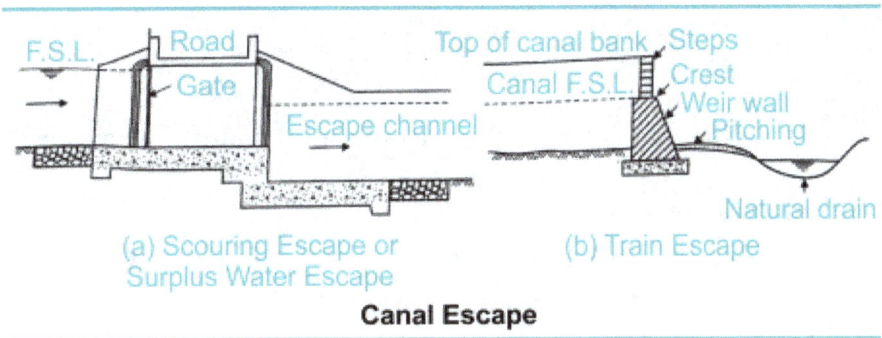

Canal Escape

Types of Canal Escape

A canal escape is an engineered structure located along an irrigation canal to allow for the controlled discharge of water.
Based on its specific function, there are three main types of escapes:

Canal Scouring Escape:
A canal scouring escape is a specialized structure strategically placed along an irrigation canal to manage and regulate the controlled release of water specifically for scouring purposes.
Its main function is to facilitate the removal of sediment and debris that may accumulate within the canal, ensuring optimal water flow and preventing potential blockages.

Surplus Escape:
A surplus escape is a crucial part of canal infrastructure designed to handle situations where there is an excess supply of water within the irrigation canal.

This escape functions as a controlled outlet, allowing the discharge of surplus water to prevent over topping of canal banks or potential damage.

Tail Escape:

Tail escapes are vital structures located at the downstream end of an irrigation canal, serving as the final point of controlled water release.

Positioned to prevent overflows and potential damage to the canal banks, tail escapes efficiently manage the water flow as it exits the canal.

19.3 CANAL OUTLET:

A canal outlet is a structure that controls the flow of water from a main canal or distributary canal into a field channel or watercourse.

It helps in distributing water to agricultural fields for irrigation purposes.

Requirements of Canal Outlet

Controlled Water Flow:
Should regulate the amount of water flowing into the fields.

Ease of Operation:
Must be easy to open, close, and adjust for farmers.

Durability:
Made from materials that can withstand environmental conditions and regular usage.

Minimal Maintenance:
Should require low maintenance and be easy to repair if necessary.

Consistent Supply:
Must provide a consistent and reliable water supply to the fields.

Prevent Erosion:
Should be designed to prevent soil erosion around the outlet.

Types of Canal Outlet

Non-Modular Outlet:

Definition: An outlet where the discharge depends on both the water level in the parent canal and the water level in the field channel.
Examples: Open sluices, Simple pipe outlets
Characteristics:
Simple design
Flow varies with changes in water levels
Inexpensive and easy to construct

Semi-Modular Outlet:

Definition: An outlet where the discharge depends only on the water level in the parent canal and is independent of the water level in the field channel, up to a certain limit.
Examples: Adjustable proportional weirs, Orifice semi-modular outlets
Characteristics:
More controlled than non-modular outlets
Flow remains steady if the canal water level is stable
Slightly more complex and costly than non-modular outlets

Rigid Module Outlet:
Definition: An outlet where the discharge is fixed and does not change with the water levels in the parent canal or the field channel.
Examples: Gibb's module, Crump's adjustable orifice
Characteristics:
Highly precise and consistent water flow
Complex design and construction
Higher initial cost but very reliable in delivering fixed quantities of water

19.4 CROSS REGULATOR:

- A cross-regulator is a structure built across a canal to control the flow of water.
- It helps manage water levels and distribute water efficiently to different parts of the irrigation system.

Requirements of Cross Regulator

Water Level Control:

- Function: Maintains the desired water level in the canal.
- Working Principle: By raising or lowering gates, the cross-regulator adjusts the water height upstream.

Flow Regulation:

- Function: Controls the amount of water flowing downstream.
- Working Principle: The gates can be opened or closed to increase or decrease the water flow as needed.

Water Distribution:

- Function: Ensures an even distribution of water to different branches of the canal system.
- Working Principle: By managing water levels, it helps divert water to various channels connected to the main canal.

Flood Control:

- ➤ Function: Helps in managing excess water during heavy rains or floods.
- ➤ Working Principle: The gates can be fully opened to allow more water to pass through, preventing overflow.

Sediment Control:

- ➤ Function: Reduces sediment buildup in the canal.
- ➤ Working Principle : By controlling the water flow speed, it helps in minimizing sediment deposition upstream.

19.5 HEAD REGULATOR:

A head regulator is a structure located at the head of a canal or distributary. It controls the flow of water from a reservoir, river, or main canal into the canal system.

Requirements of Head Regulator

Flow Control:
- ➤ *Function*: Regulates the amount of water entering the canal.
- ➤ *Working Principle:* Gates or valves can be opened or closed to control the water flow.

Water Level Management:
- ➤ *Function:* Maintains a consistent water level in the canal.
- ➤ *Working Principle:* By adjusting the gate positions, it ensures the desired water level is maintained.

Distribution Regulation:
- ➤ *Function:* Directs water into multiple canals or branches.
- ➤ *Working Principle*: The head regulator can have multiple openings leading to different canals, each controlled separately.

Sediment Control:
- ➤ *Function:* Prevents excessive sediment from entering the canal system.
- ➤ *Working Principle:* Designed to settle out sediments before water enters the canal, or by flushing out sediments through special gates.

Safety and Flood Control:
Function: Prevents overflow and damage during high water levels or floods.
Working Principle: Can release excess water safely to manage high inflows and protect the canal structure.

Measurement and Monitoring:
Function: Helps in measuring and monitoring water flow for efficient management.
Working Principle: Often equipped with measuring devices to monitor the amount of water being discharged.

www.ingramcontent.com/pod-product-compliance
Lightning Source LLC
Chambersburg PA
CBHW082238220526
45479CB00005B/1276